MW00447232

STUDENT GUIDE

for

MACHINE SHOP OPERATIONS AND SETUPS

Prepared by
Howard M. Draves

Based on the Textbook
MACHINE SHOP OPERATIONS AND SETUPS
Fourth Edition
by Porter - Lascoe - Nelson
Published by
American Technical Publishers, Inc.
Homewood, Illinois 60430

© 1967, 1973 by American Technical Publishers, Inc.
All rights reserved

4 5 6 7 8 9 – 73 – 28 27 26 25

Printed in the United States of America

ISBN 978-0-8269-1844-4

STUDY GUIDE FOR MACHINE SHOP OPERATIONS AND SETUPS

HOW TO USE THIS STUDY GUIDE FOR EFFECTIVE LEARNING

The following explanation tells you what is in your Study Guide and how it will help you learn most rapidly and effectively.

WHAT IS IN THE STUDY GUIDE

This Study Guide is arranged in ten sections, numbered from 1 to 10. Each section contains a number of short Progress Quizzes and one EXAMINATION.

The Progress Quizzes help you to check quickly what you have learned from the study assignments in the textbook and to indicate quickly any point needing further study. In the Progress Quiz the reference page number in the textbook is printed directly after the question. Correct answers to questions in the Progress Quizzes are in the ANSWER KEY in the back of the Study Guide.

A MEMORY JOGGER is included in Sections 3, 6, and 9, preceding the examination. This is a review quiz on preceding chapters. The answers to the Memory Jogger questions are in the Answer Key. Directly after the questions in the Memory Jogger are textbook page references.

The first page of each section tells you the chapter or chapters of the textbook covered in that section, the pages to study in the textbook in preparation for each Progress Quiz, and lists Progress Quizzes and Examination in the proper order. Your Aim for the section is explained. The comments under the heading, SOME SUGGESTIONS will help you make good use of your study time.

IMPORTANT On the following pages is a combined Study Assignment Check List and Progress Record, which provides a complete list of the 27 Progress Quizzes, 3 Memory Joggers, and 10 Examinations in the Study Guide. A Date Completed column has been provided for recording your study progress.

Be sure to read the directions on HOW TO ANSWER EACH TYPE OF QUESTION and ANALYZING YOUR RETURNED EXAMINATIONS which are on the pages that follow the Check List.

STUDY ASSIGNMENT and PROGRESS RECORD

SECTION 1 Chapters 1 and 2 of the textbook, covering MACHINE TOOLS as the basis of all industry, and MEASURING TOOLS (precision and semi-precision)

	Study Guide Page Number	Date Assigned	Date Completed
Progress Quiz 1	3-4		
Progress Quiz 2	5-6		
Progress Quiz 3	9		
Progress Quiz 4	23-24		
Progress Quiz 5	25-26		
EXAMINATION 1	27-31		

SECTION 2 Chapters 3 and 4 of the textbook, covering BENCH TOOLS, LAYOUT TOOLS, and POWER SAWS

	Study Guide Page Number	Date Assigned	Date Completed
Progress Quiz 6	35-36		
Progress Quiz 7	37-38		
Progress Quiz 8	39-40		
EXAMINATION 2	41-44		

SECTION 3 Chapter 5 of the textbook, covering DRILL PRESSES (types, set-ups, and operations)

	Study Guide Page Number	Date Assigned	Date Completed
Progress Quiz 9	47-48		
Progress Quiz 10	49-50		
Progress Quiz 11	51-52		
MEMORY JOGGER 1	53-54		
EXAMINATION 3	55-58		

SECTION 4 Chapters 6, and part of Chapter 7, covering ENGINE LATHES (types, accessories, attachments, and external operations)

	Study Guide Page Number	Date Assigned	Date Completed
Progress Quiz 12	61-62		
Progress Quiz 13	63-64		
Progress Quiz 14	65-66		
EXAMINATION 4	67-71		

SECTION 5 Chapter 7 (last half) and Chapter 11 covering TURRET LATHES and AUTOMATIC SCREW MACHINES (types, setups, and operations)

	Study Guide Page Number	Date Assigned	Date Completed
Progress Quiz 15	75-76	________	________
Progress Quiz 16	77-78	________	________
Progress Quiz 17	79-80	________	________
EXAMINATION 5	81-85	________	________

SECTION 6 Chapter 8 of the textbook, covering SHAPERS (types, setups, and operations)

Progress Quiz 18	89-90	________	________
MEMORY JOGGER 2	91-92	________	________
EXAMINATION 6	93-96	________	________

SECTION 7 Chapters 9 and 10 of the textbook, covering MILLING MACHINES (types, accessories, setups, and operations)

Progress Quiz 19	99-100	________	________
Progress Quiz 20	101-102	________	________
Progress Quiz 21	103-104	________	________
EXAMINATION 7	105-110	________	________

SECTION 8 Chapter 12 of the textbook, covering GRINDING MACHINES (types, setups, and operations)

Progress Quiz 22	113-114	________	________
EXAMINATION 8	115-117	________	________

SECTION 9 Chapters 13 and 14 of the textbook, covering STEEL AND ITS ALLOYS and HEAT TREATING OF STEEL

	Study Guide Page Number	Date Assigned	Date Completed
Progress Quiz 23	121-122	________	________
Progress Quiz 24	123-124	________	________
MEMORY JOGGER 3	125-126	________	________
EXAMINATION 9	127-130	________	________

SECTION 10 Chapters 15, 16, and 17 of the textbook, covering MATERIAL MACHINABILITY, NUMERICAL CONTROL, and ELECTRICAL ENERGY MACHINING PROCESSES

	Study Guide Page Number	Date Assigned	Date Completed
Progress Quiz 25	133-134	________	________
Progress Quiz 26	137	________	________
Progress Quiz 27	139-140	________	________
EXAMINATION 10	141-143	________	________

HOW TO ANSWER EACH TYPE OF QUESTION in this Study Guide

There are three different types of questions in the Progress Quizzes and Examinations: 1. AGREE—DISAGREE 2. COMPLETION 3. MULTIPLE-CHOICE. Here is an example of each type of question. Study each one carefully. These examples do not appear on the examinations or Progress Quizzes.

AGREE—DISAGREE

1. (A) D The smallest division on the scale of a steel rule is 1/64 inch.

If you AGREE with the statement, draw a circle around the letter A.
If you DISAGREE with the statement, circle the letter D.

COMPLETION

2. A fluted tool used to cut internal threads is called a tap.

Complete the statement by writing or printing on the blank line, the word or words needed to make a complete and correct statement.

MULTIPLE-CHOICE

3. A tapping attachment must be used to power-tap holes on a—

a. radial drill press
b. vertical or upright drill press
(c.) sensitive drill press

In multiple-choice questions there are three or four phrases from which to choose one that makes an accurate, complete statement when added to the incomplete sentence. Draw a circle around the letter in front of your choice, as shown in the example.

OTHER TYPES OF QUESTIONS: Where there are other types of questions in the examinations, directions for answering them are given. Read directions carefully.

HOW ANALYZING A RETURNED, CORRECTED, AND GRADED EXAMINATION CAN HELP YOU

When a corrected and graded examination is returned to you, more can be learned from it than just the questions you answered incorrectly! You can help yourself if you take a few minutes to ANALYZE each corrected examination when it is returned to you.

ANALYZING these questions means asking yourself the questions listed below (and others you may think of). This will help you develop good study habits.

1. Are you using the Progress Quiz questions to help check how thoroughly and how much you have learned?
2. Do you review the textbook on any Progress Quiz question you answer incorrectly?
3. Do you review the Progress Quiz questions and answers before taking an examination?
4. Do you read the Progress Quiz and Examination questions carefully, to be sure you understand them?
5. Does it appear that you sometimes answer questions before taking time to think about them?
6. Do you find that you did not read the captions on the illustrations carefully enough and missed important points in this manner?
7. Do you study the illustrations carefully at the time the text mentions them by Figure number?
8. Do you take time to refer to your dictionary when you read new words in the textbook?
9. As you study, do you look for the basic principle which underlies the textbook explanations and keep this principle in mind?
10. Do you have difficulty with any particular type of question? If so, have you tried to find out why?

If you will use these suggested check points to analyze your returned examinations, you will learn more and gain a better understanding of the subject.

NOW — turn to the first page of SECTION 1 and begin your study as directed there.

SECTION 1 — Study Guide for MACHINE SHOP OPERATIONS AND SETUPS

TEXTBOOK CHAPTERS COVERED

Chapter 1 — Machine Tools: Measure of Man's Progress (Textbook pages 1 to 9)
Chapter 2 — Measuring Tools: Semi-Precision and Precision (Textbook pages 9 to 45)

CONTENTS OF SECTION 1

Assignment 1 and Progress Quiz 1 — Study pages 1 to 9. Importance of machine tools to man's progress; safety rules and regulations as outlined by OSHA.

Assignment 2 and Progress Quiz 2 — Study pages 9 to 22. Use of semi-precision measuring tools.

Assignment 3 and Progress Quiz 3 — Study pages 22 to 26. Decimal and metric scales of measurement and how to convert measurements from one scale to the other.

Assignment 4 and Progress Quiz 4 — Study pages 26 to 37. Reading of micrometer calipers, inside micrometer calipers, inside micrometers, micrometer depth gage, and the vernier micrometer.

Assignment 5 and Progress Quiz 5 — Study pages 37 to 49. Reading of the metric micrometer, the vernier caliper and height gage, the vernier bevel protractor, and the dial indicator.

EXAMINATION 1 — Based on textbook pages 1 through 49 in the Assignments for SECTION 1 of this STUDY GUIDE

WHAT THE STUDY OF SECTION 1 WILL DO FOR YOU

The assignments in Section 1 are presented in logical learning order. For example, in Assignment 1 you will learn first how important and basic the machine tool is to man's progress and the economy of the country. Without improvements in tool designs and in increased knowledge of machining methods, many of our modern conveniences we accept as commonplace would not exist. After this introduction, you will learn what knowledge and work are expected of you as a general Machinist or as a Machinist-Technician. In Assignments 2 to 5 you will read and learn what Semi-Precision and Precision Measuring Tools are used by machinists and how they care for such tools. Accuracy in machine tool operation when machining parts is not possible, without first knowing how to take accurate measurements. Study the material in Section 1 carefully.

MACHINE SHOP OPERATIONS AND SETUPS

ASSIGNMENT 1 and PROGRESS QUIZ 1

SUBJECT The Importance of the Machine Tool to Man's Progress, Skills Required by Machinist-Technicians, the Observance of Clean, Safe Work Habits

TEXTBOOK PAGES TO STUDY 1 to 9

"YOUR AIM" for this assignment is to learn or develop a keen sense of pride in what you are trying to accomplish in the entire study of this textbook. In machine shop work every part must not only be machined accurately to size but also must have a fine surface finish. Without pride in work accomplishment, the machinist cannot produce the quality of work required. To help in this respect, you will learn what skills are required of a machinist in the machine shop and what safe working habits to observe in the performance of shop work.

SOME SUGGESTIONS

To prepare you for the study of later assignments, keep in mind that because of the precise nature of work performed in the shop you must learn to do all work in a skillful and competent manner. Do not forget that safe practices must be observed at all times to prevent serious injury to yourself or fellow workers.

PROGRESS QUIZ 1

THE PURPOSE OF THIS QUIZ: To help you determine whether or not you can recall the important points studied in this assignment

1. The one factor which has contributed most to man's material well-being and progress has been governed by the ______________ he has developed. (See textbook, page 1.)

2. The three basic geometric forms from which all product designs are developed are the ______________, the ______________, and the cylinder. (Page 3, Fig. 2)

3. A D Only those machines which use cutting tools to shape or machine parts to size can be classified as machine tools. (Page 4)

4. Accuracy of ____________________ has been the one factor which has contributed most to the preciseness of operations which machine tools are capable of today. (Page 6)

(OVER)

5. Hand tools such as ____________ and scrapers are unsafe to use unless they are equipped with wood handles. (Page 7)

6. As a safety precaution, ______________ ______________ must be worn at all times when operating any machine tool. (Page 8)

7. When operating machine tools, do not wear jewelry items such as ________. (Page 8)

8. When returning tools to the tool crib or storage room, report any tools which may have been ________________. (Page 8)

9. To prevent falls, cover all oil spots on floors with an ________-__________ compound. (Page 8)

(END OF QUIZ)

SUGGESTION: Check your answers to these questions with the Answer Key in the back of this Study Guide. Review any point incorrectly answered before you proceed with the study of Assignment 2.

MACHINE SHOP OPERATIONS AND SETUPS

ASSIGNMENT 2 AND PROGRESS QUIZ 2

SUBJECT Semi-Precision Measuring

TEXTBOOK PAGES TO STUDY 9 to 22

"YOUR AIM" As pointed out in Assignment 1, accurate work is entirely dependent upon the machinist's ability to take accurate measurements. In this assignment you will read and learn about the various semi-precision tools used by the machinist in measuring parts which require no greater accuracy than 1/64 inch.

SOME SUGGESTIONS

Most of the tools discussed in this assignment are related to an understanding of the basic line-graduated measuring tool called a steel rule. There are many variations of this rule to serve specific needs so learn to select the one most proper for the job. On steel rules there are four different scales, ranging from the smallest division of 1/64 to 1/8 inch. Study figs. 1 and 2 on page 10 of the textbook so that you understand and can identify the divisions on each scale of the rule. Note in the use of inside and outside calipers that work is measured by contact and that all such measurements of work must then be transferred to the steel rule, to determine the size of parts. Figs. 22 and 25, on pages 19 and 21 of the textbook show you how this should be done to assure an accurate reading which can result only when one caliper leg is flush or even with the end of the rule. Observe this carefully. Now answer questions in Progress Quiz 2.

PROGRESS QUIZ 2

THE PURPOSE OF THIS QUIZ: To help you determine how well you have learned the correct use of all measuring tools covered in this assignment

1. The smallest division on the scale of a steel rule is__________inch. (See textbook, page 9.)

2. A D The hook rule is preferred to the slide caliper rule when measuring the diameter of round stock. (Page 15)

3. The combination set is used chiefly for________________work in the machine shop. (Page 16)

4. The protractor head on a combination set is designed to measure or check ____________________. (Page 16)

(OVER)

5. A measuring tool designed especially for depth measurement of holes, slots, and recesses is called a_______________ _______________gage. (Page 17)

6. Tools used to measure work by contact are called_______________. (Page 19)

7. A D To assure an accurate measurement using inside or outside calipers, the calipers must be held lightly in the finger tips. (Page 19)

8. To obtain an accurate measurement using_______________, the machinist must develop a keen sense of touch. (Page 21)

9. A D Inside caliper measurements must be checked with a steel rule or a micrometer. (Page 21)

(END OF QUIZ)

SUGGESTION: Check your answers to these questions with the Answer Key in the back of this Study Guide. Review any point incorrectly answered before you proceed with the study of Assignment 3.

MACHINE SHOP OPERATIONS AND SETUPS

ASSIGNMENT 3 AND PROGRESS QUIZ 3

SUBJECT Decimal and Metric Systems of Measurement

TEXTBOOK PAGES TO STUDY 22 to 26

"YOUR AIM" for this assignment will be to develop a complete understanding of the Decimal and Metric systems of measurement. You will want to learn the meaning of a decimal, how a decimal value is read, and, most of all, how to convert fractional dimensions to decimal values. Because of expanding world trade, it is also necessary for the machinist to be familiar with the European Metric system of measurement.

SOME SUGGESTIONS

When studying this assignment, keep in mind that the precision measuring tools commonly used by machinists and discussed in the following two assignments are capable of providing measurements accurate only to four decimal places. Most shop work falls in this category. Any closer accuracy than four decimal places can be checked only through the use of special air or electronic gages. Concentrate on learning the correct way to read and write the three- and four-place decimal values. On page 25 of the textbook you are provided a table showing the decimal equivalents for all common fractions. This table is important because the machinist must frequently convert a drawing dimension in fractions to decimal form so he can measure his work with a micrometer. Charts are available in most shops for reference, but to save time it would be wise to memorize the decimal equivalents of the following common fractions: 1/64, 1/32, 1/16, 1/8, 3/16, 1/4, 5/16, 3/8, 1/2, 9/16, 5/8, 3/4, and 7/8. These are the fractions which occur most frequently in shop work. The others are less common and can always be found by referring to a shop chart. When studying the metric scale explained on pages 25 and 26 of the textbook, remember that the basic unit of length in this system is the meter. To become familiar with this system, you must learn the prefixes and abbreviations assigned to the meter for the smaller units of measurements such as the millimeter, the centimeter, etc. So that you will be able to convert metric dimensions to the decimal system, study the conversion Table II on page 26 of the textbook.

Now proceed to answer the questions in Progress Quiz 3 which follows.

PROGRESS QUIZ 3

THE PURPOSE OF THIS QUIZ: To test your understanding of the decimal and metric scales of measurement and how to convert measurements from one scale to the other

1. A D When piece parts must be measured to a greater accuracy than 1/64 inch, the decimal system of measurement is used. (See textbook, page 22.)

2. A D A piece part dimensioned on a blueprint in decimal form indicates to the machinist that this part must be machined closer to size than if it were dimensioned as a fraction. (Page 22)

3. In the decimal system of measurement, fifty-five thousandths of an inch would be written as_______________. (Page 23)

4. In the decimal system of measurement, ten thousandths of an inch should be written as_____________. (Page 23)

5. In the decimal system of measurement, five ten-thousandths of an inch should be written as_____________. (Page 24)

6. The decimal equivalent of 1/16 inch is _____________. (Page 25)

7. The decimal equivalent of 3/8 inch is _____________. (Page 25)

8. The decimal equivalent of 5/8 inch is _____________. (Page 25)

9. The metric system of measurement is based on a length called the _______________. (Page 25)

10. A D In the metric system of measurement, 1 millimeter is the same as .03937 in the decimal system. (Page 26)

(END OF QUIZ)

SUGGESTION: Check your answers to these questions with the Answer Key in the back of this Study Guide. Review any point incorrectly answered before you proceed with the study of Assignment 4.

MACHINE SHOP OPERATIONS AND SETUPS

ASSIGNMENT 4 AND PROGRESS QUIZ 4

SUBJECT Precision Measuring Tools Including the Micrometer Caliper, the Inside Micrometer Caliper, the Inside Micrometer, the Micrometer Depth Gage and the Vernier Micrometer

TEXTBOOK PAGES TO STUDY 26 to 36

"YOUR AIM" for this assignment will be to learn how to apply what you read and learned in Assignment 3 about the decimal system of measurement to the actual reading and use of all precision measuring tools listed under the heading of "SUBJECT" above.

SOME SUGGESTIONS

To assure a more complete understanding of how each one of the precision measuring tools discussed in this and the next assignment are to be read, be sure to study the supplementary information beginning on page 13 of this Study Guide. Now you should be well prepared to answer all questions in Progress Quiz 4, following the supplementary information.

BE SURE —

TO READ THIS

HELPFUL ADDITIONAL

INSTRUCTION

ON

PRECISION MEASURING

INSTRUMENTS

How To Read a Plain Micrometer

First learn the correct values for each line or division on the barrel and sleeve, or thimble, of the micrometer.

Each line on the barrel equals .025 or twenty-five thousands of an inch. Note that every fourth line is numbered 1, 2, 3, 4, etc. These figures represent hundreds of thousands, or .100, .200, .300, .400, etc. The numbering of every fourth line on the barrel is for fast reading and eliminates the necessity of counting each of the .025 lines showing on the barrel. For example, in the reading shown, note that the last numbered line showing on the barrel closest to the end of the sleeve, or thimble, is 4 so we know the barrel reading is .400.

Next, count the number of lines showing between the number 4 line on the barrel and the end of the thimble. In the reading shown, there are two lines showing so add 2x.025 or .050 to the first reading of .400. Now we have .400 + .050 or .450.

Each line on the sleeve, or thimble, equals .001. To find this sleeve or thimble reading, simply take the value of the thimble line which matches the horizontal index line on the barrel. In the example shown, the number 12 line is the matching line so add .012 to the previous reading. The total reading is .400 + .050 + .012 or .462.

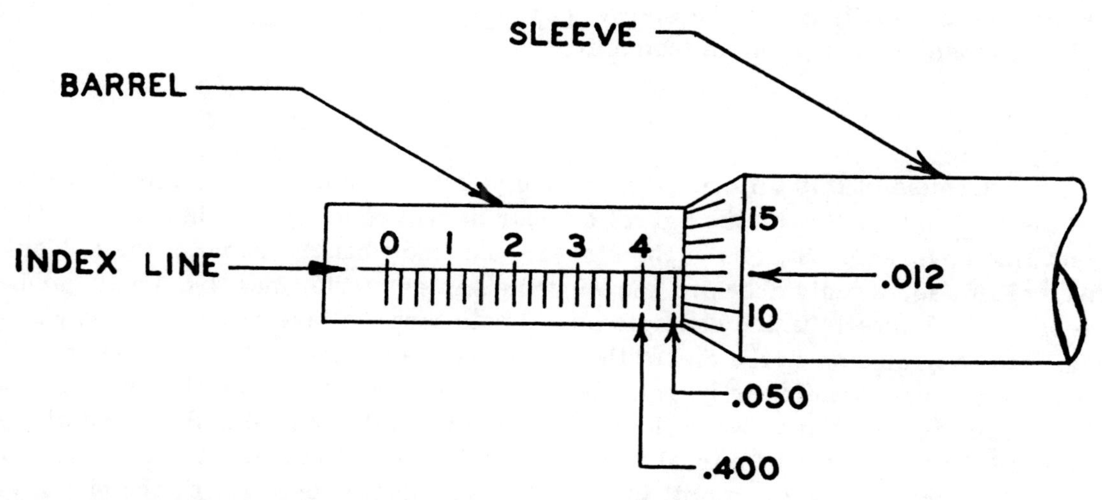

How To Read an Inside Micrometer Caliper and a Micrometer Depth Gage

NOTE: The reading of an inside micrometer caliper and a micrometer depth gage is the same, so the following explanation applies to both instruments even though the example shown will be that of a micrometer depth gage.

Each division or line on the barrel represents .025; every fourth line is numbered as on a plain micrometer, and these numbers represent hundred thousands, or .700, .800, .900, etc.

Each line on the sleeve or thimble represents one thousands of an inch or .001, .002, .003, .004, etc.

Note when reading these instruments that the same general procedure applies as used on a plain micrometer only that the scales on the barrel and sleeve are laid out just the opposite of those on a plain micrometer. Those on the barrel increase in value from right to left instead of from left to right as on a plain micrometer, and those on the sleeve increase in value counterclockwise instead of clockwise as on a plain micrometer.

Because of this difference in scales you must remember to use the value of the next hidden line on the barrel as your barrel reading; in other words, the last line covered by the end of the sleeve. To find what the value of the hidden barrel line is, simply check to see what the values of the lines are which can be seen on the barrel. In the example shown note that the last numbered line which can be seen, closest to the end of the sleeve is the number 7 line, so you know that your barrel reading is going to be .600 plus. Now note that the two lines on the barrel are visible between the number 7 line on the barrel and the end of the sleeve. Each line on the barrel represents .025, two lines equal .025x2 or .050. If the last visible line on barrel is .050 then the next hidden line or the one covered by the sleeve must be .025. Now add this .025 to your hidden numbered line which is .600. Your total barrel reading is then .600 + .025 = .625.

How To Read an Inside Micrometer Caliper and a Micrometer Depth Gage

To obtain your sleeve or thimble reading, simply take the value of the sleeve line which matches the horizontal index line on the barrel. In this case the number 19 line is the matching line so add .019 to your barrel reading for a total reading of .625 + .019 = .644.

If extension rods are used with either of these instruments as is often the case, remember to add the length of the rod to your instrument reading. For example, if in the reading shown, a 3-inch-long rod was being used you would add only two inches to the measurement of .644 and your total reading would be 2.644.

The rod is 3 inches long but only 2 inches enter into the measurement, as 1 inch of the rod length is used in fastening the rod into the head of the micrometer.

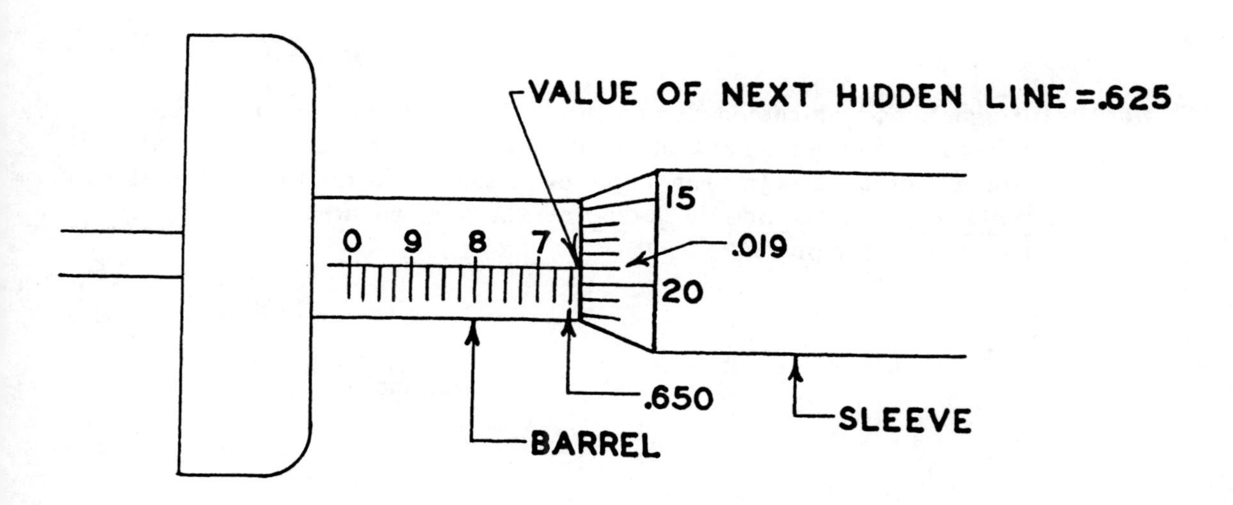

TOTAL READING =.625+.019=.644

How To Read a Vernier Micrometer

The barrel and sleeve of a vernier micrometer are read the same as the barrel and sleeve of a plain micrometer. All graduations have the same value as those on the barrel and sleeve of a plain micrometer. The only difference in reading occurs when a line on the sleeve does not exactly match the horizontal index line on the barrel. See example shown, and note that the index line on barrel falls between the 7 or 8 line on the sleeve. When this occurs always take the value of the line which is just below the index line of the barrel for your sleeve reading. In this example the line below the index line of the barrel is the number 7 line so your sleeve reading would be .007. To find how much greater your reading on the sleeve is over .007 you must refer to the vernier scale.

Each line on the vernier scale is numbered from 1 to 9 and these numbers represent values in one ten thousands of an inch or .0001, .0002, .0003, .0004, etc. To read this scale, simply take the value of the line on vernier scale which matches a line on the sleeve. In the example shown, the number 3 line is the matching line so add .0003 to your barrel and sleeve reading for your total reading. Total reading = .350 + .007 + .0003 = .3573

NOTE: It should be understood that when any line on the sleeve matches the index line on the barrel that the vernier scale reading will be zero. For example, if in the reading shown, the number 7 line on the sleeve matched the index line on the barrel the total reading would simply be .350 + .007 or .357.

(example on next page)

How To Read a Vernier Micrometer

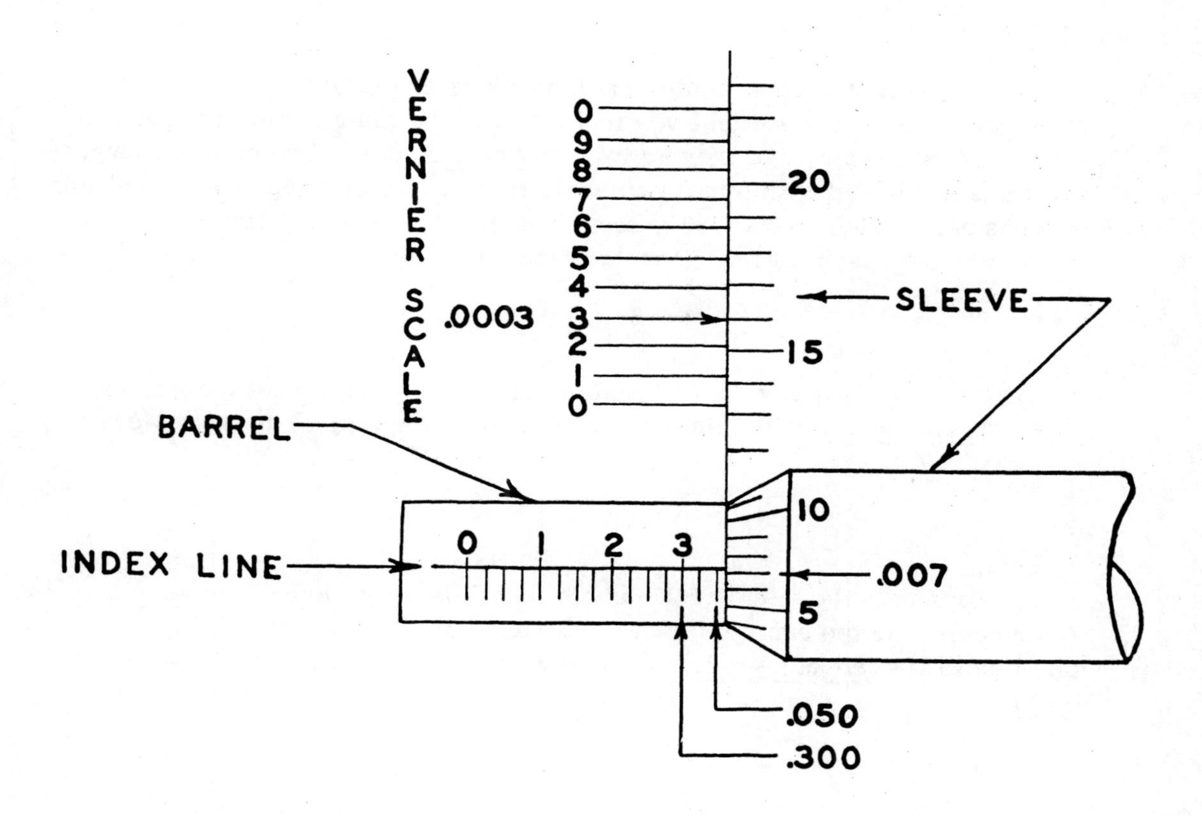

TOTAL READING=.300+.050+.007+.0003=.3573

How To Read a Vernier Caliper

First learn the correct values for each line or division on the top or main scale.

Note that the top or main scale divisions are indicated by three different length lines. The longest vertical lines on the main scale are numbered with large numerals, and they represent whole inches. The medium length lines are numbered with small numerals from 1 to 9 and represent hundreds of thousands—.100, .200, .300, etc. The shortest vertical lines on the main scale each represent twenty-five thousands or .025.

To read the top, or main scale, keep in mind that the zero line on the lower vernier scale determines your reading of the top, or main scale.

First note the value of the whole inch line to the left of the zero line on lower scale. In the example shown, the number 6 line of top scale is the first inch line to the left of zero line on lower scale so our whole inch reading is 6.000.

Next, count the number of lines showing between the six-inch line of the top scale and the zero line of the lower scale. In this example there are two lines so add 2 x .025 or .050 to the whole inch reading of 6.000. 6.000 + .050 = 6.050.

To read the lower vernier scale remember that each line on this scale equals .001, or one thousands of an inch. To obtain this reading, simply take the value of the vernier scale line which matches a line on the top scale. In this case the number 10 line is the matching line so add .010 to the top scale reading. 6.050 + .010 equal 6.060.

(continued on next page)

How To Read a Vernier Caliper

If the zero line on the vernier scale matches a line on the top scale, the vernier scale reading will be zero. In this case you simply take the value of the line on top scale which matches the zero line on lower scale. For example, if in the reading shown, the zero line on lower scale matched the number 1 line on top scale, the total reading would be 6.000 + .100 = 6.100.

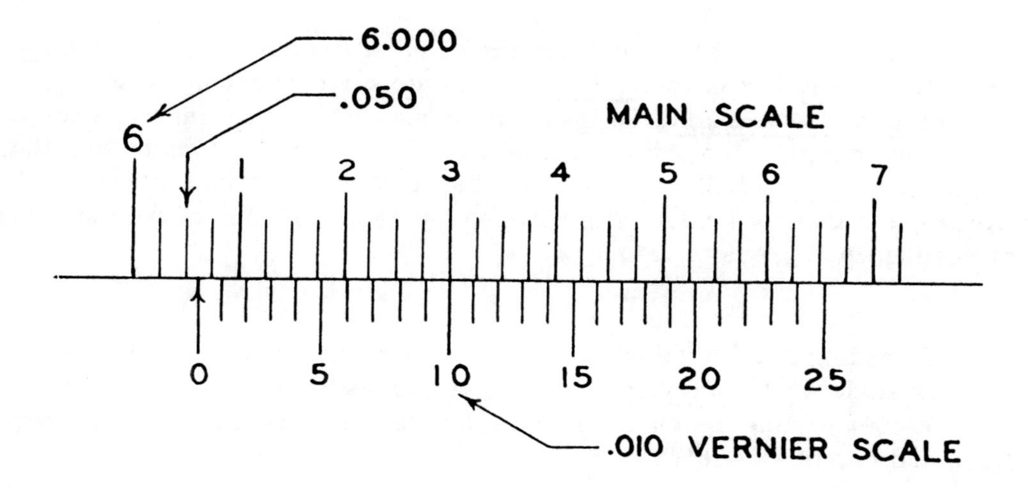

TOTAL READING = 6.000 + .050 + .010 = 6.060

How To Read a Vernier Bevel Protractor

First learn the correct values of each line or division on the top or degree scale and the lower or vernier scale.

Each line on top scale represents 1 degree.

Each line on lower vernier scale represents 5 minutes.

To read the top scale, first note whether the value of lines on the top scale increases from right to left of the zero line on lower scale or from left to right of the zero line on lower scale. In the example shown the lines increase in value from left to right. To obtain your reading in degrees remember that the zero line on lower scale indicates your reading in degrees of the top scale. In this example shown, the zero falls between the 18 and 19 line of the top scale so your reading in degrees is 18°.

To find your reading in minutes on the lower scale, simply take the value of line on lower scale which matches a line on the top scale. In this example shown the number 45 line matches a line on the top scale so your total reading in degrees and minutes is 18° 45'.

NOTE: There can only be one line on lower scale which will match a line on the top scale. If and when the zero line on lower scale matches a line on the top scale, your total reading will be found by taking the value of the top scale line which matches with the zero line on lower scale. If for example in the reading shown, the zero line matched the number 18 line on top scale, your total reading would be 18° and no minutes.

(example on next page)

How To Read a Vernier Bevel Protractor

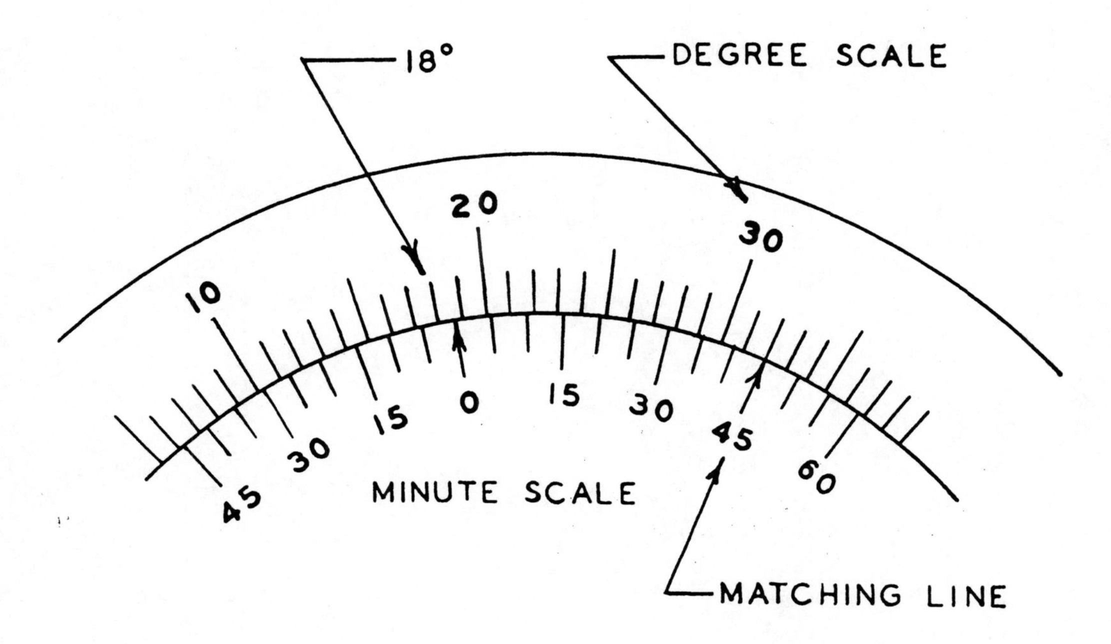

TOTAL READING = 18°45′

PROGRESS QUIZ 4

THE PURPOSE OF THIS QUIZ: To test your understanding of how to read and use the precision measuring tools discussed in this assignment

1. A D Each vertical cross line on the barrel of a micrometer caliper represents .025 inch.. (Page 27)

2. A D Each line or graduation on the thimble of a micrometer caliper represents .001 inch. (Page 27)

Directions: Read the micrometer caliper settings illustrated. Write the correct reading on the line under each illustration.

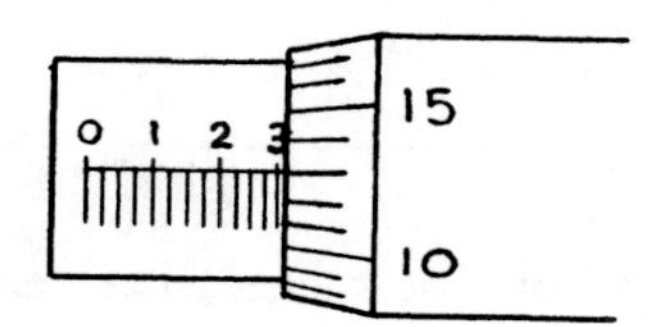

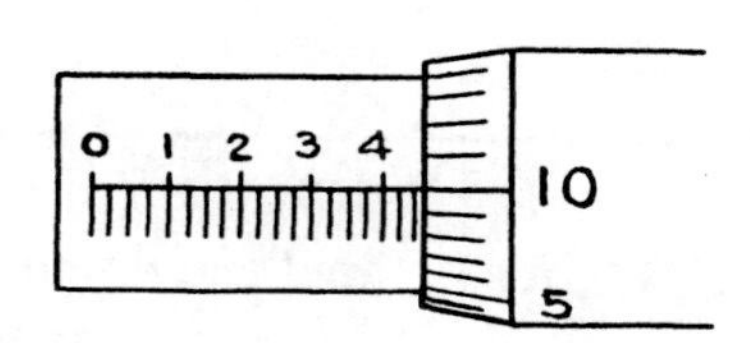

3. ______________________ 4. ______________________

(Pages 28-29)

Directions: Read the Inside Micrometer Caliper settings illustrated. Write the correct reading on the line under each illustration.

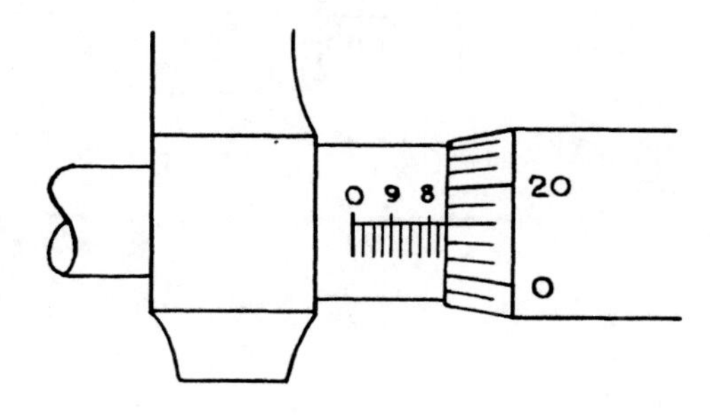

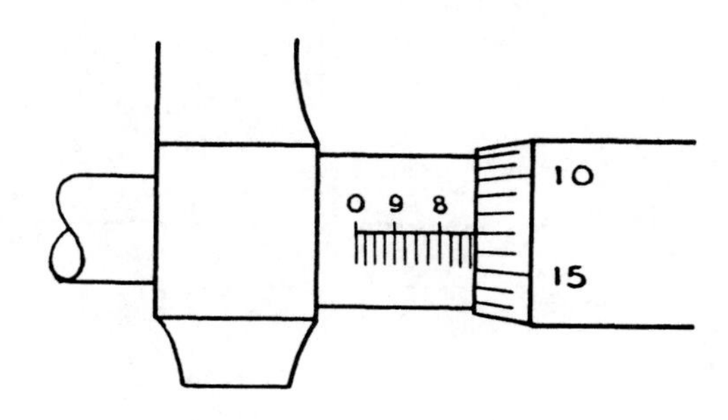

5. ______________________ 6. ______________________

(Pages 31-32)

(OVER)

Directions: Assuming that no extension rods are being used, read the micr meter depth gage settings illustrated. Write your answer on the line under each illustration.

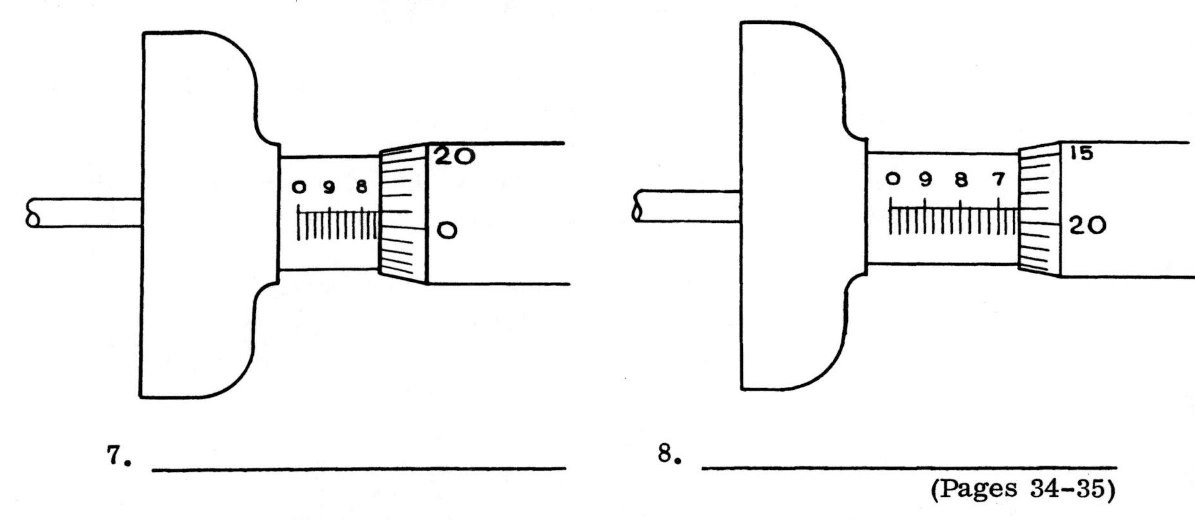

7. ____________________ 8. ____________________

(Pages 34-35)

Directions: Read the vernier micrometer caliper settings illustrated. Write your answers on the lines under each illustration.

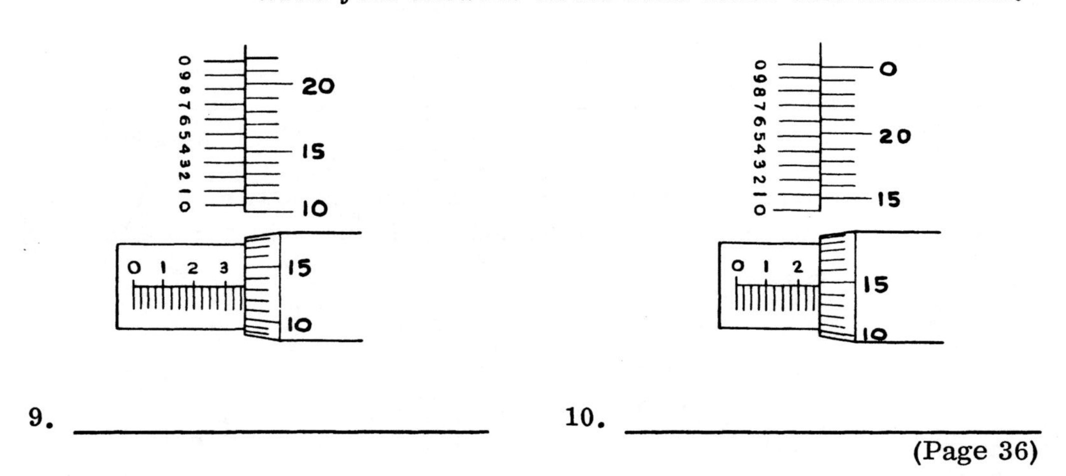

9. ____________________ 10. ____________________

(Page 36)

(END OF QUIZ)

SUGGESTION: Check your answers with the Answer Key in the back of the Study Guide. Review the study of textbook and the supplementary information in the Study Guide on any points missed, then proceed with the study of Assignment 5.

ASSIGNMENT 5 AND PROGRESS QUIZ 5

SUBJECT Precision Measuring Tools Including the Vernier Micrometer, the Metric Micrometer, the Vernier Caliper, the Vernier Height Gage, the Vernier Bevel Protractor, and the Dial Indicator

TEXTBOOK PAGES TO STUDY 36 to 49

"YOUR AIM" in this assignment will be to learn the correct use and reading of the precision instruments listed under the heading of "SUBJECT" above.

SOME SUGGESTIONS

To learn how each of these precision measuring tools is read, it is wise to learn first the correct value of each line or division on each scale of the instrument. When you learn each value, you should concentrate next on learning what is used to determine the reading of each scale. The final or total reading of the instrument in each case then is simply a matter of finding the total sum of all scale readings by addition. If you follow these three steps in the order stated, you will experience no problem with the reading of any precision instrument.

PROGRESS QUIZ 5

THE PURPOSE OF THIS QUIZ: To test your total understanding of how to read and use the precision measuring tools discussed in this final assignment of Section 1

1. A D The vernier micrometer is the only micrometer which can be used to measure a part dimensioned to four decimal places. (See textbook, page 36)

2. A D Each vertical line on the barrel of a metric micrometer represents .5MM. (Page 36)

3. Each line or division on the thimble of a metric micrometer represents _____________ millimeter. (Page 36)

4. Each line on the true or main scale of a vernier caliper represents _____________ inches. (Pages 38-39)

5. A D On a vernier caliper the true or main scale reading is determined by the zero mark on the vernier scale. (Page 39)

(OVER)

6. A D The vernier height gage and the vernier caliper readings are obtained in the same manner. (Page 43)

7. A precision tool used in measuring angles to an accuracy of five minutes is called a ____________________bevel________________. (Page 44)

Directions: Read the vernier bevel protractor setting illustrated, then write your answer for the correct reading on the line provided.

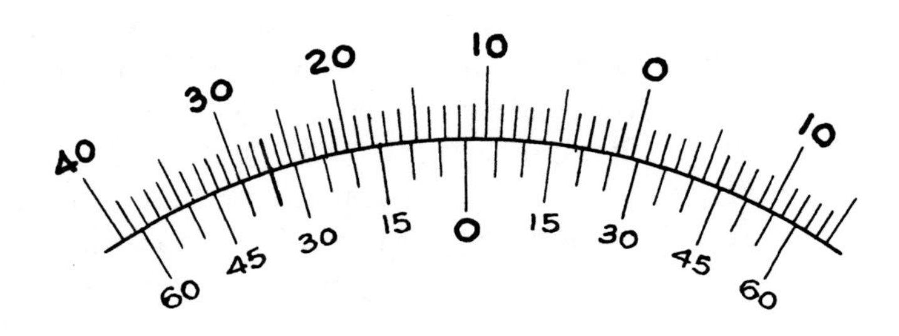

8. ______________________________ (Page 45)

(END OF QUIZ)

SUGGESTION: Check all answers to quiz questions by referring to Answer Key in the back of the Study Guide. This completes all of your study for Section 1. Now answer all questions in Examination 1 of Section 1 beginning on the next page.

CUT OFF HERE

MACHINE SHOP OPERATIONS AND SETUPS

TOPIC: Measuring Tools

Examination 1

Based on pages 1 to 49 of the textbook

Student's Name ______________________ Student Number ____________

Street ______________ City ______________ State __________ Zip Code ______

1. The number of $\frac{1}{16}$ -inch divisions in $\frac{1}{2}$ inch on the scale of a steel rule is—

   a. four
   b. eight
   c. six

2. The smallest division on the scale of a steel rule is—

   a. $\frac{1}{64}$ inch
   b. $\frac{1}{8}$ inch
   c. $\frac{1}{32}$ inch

3. The amount of tolerance generally permitted on all scale measurements is plus or minus—

   a. .0625
   b. .03125
   c. .0156

4. The best rule to use when measuring the diameter of round stock is the—

   a. plain steel rule
   b. hook rule
   c. slide caliper rule

5. The ____________ ______________ of a combination set is frequently used to lay out centerlines on the end of round stock.

6. A D Outside calipers can be used to measure the diameter of holes and round stock.

F4-3

(OVER)

7. On the lines provided, write the decimal equivalent of each of the following fractions.

$\frac{1}{16}$ ____________ $\frac{1}{32}$ ____________ $\frac{1}{8}$ ____________

$\frac{1}{4}$ ____________ $\frac{1}{2}$ ____________ $\frac{7}{16}$ ____________

8. In the metric system of measurement, one one-hundreth meter is a—

   a. decimeter
   b. millimeter
   c. centimeter

9. A D On a steel rule graduated on a metric scale, each small line or division equals 1 millimeter.

10. To convert a millimeter dimension to a decimal dimension, you must multiply the number of millimeters by—

   a. .03937
   b. .3937
   c. .3927

Directions: Study the micrometer caliper readings then write the correct readings on the line provided under each illustration.

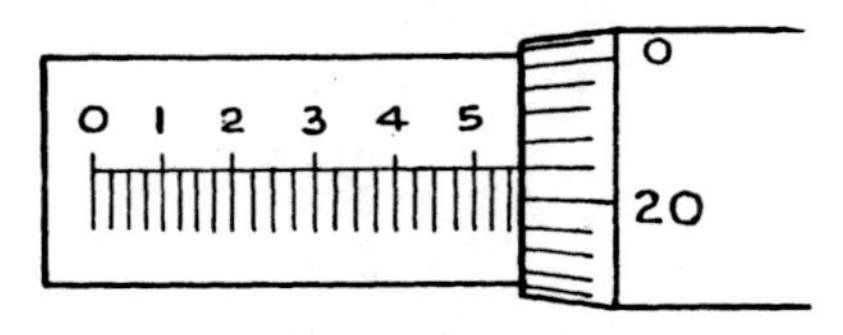

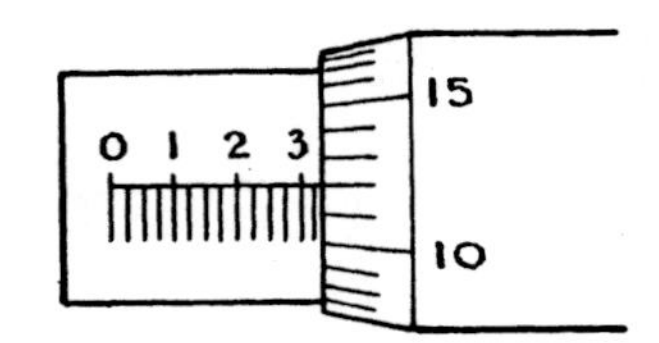

11. ____________________ 12. ____________________

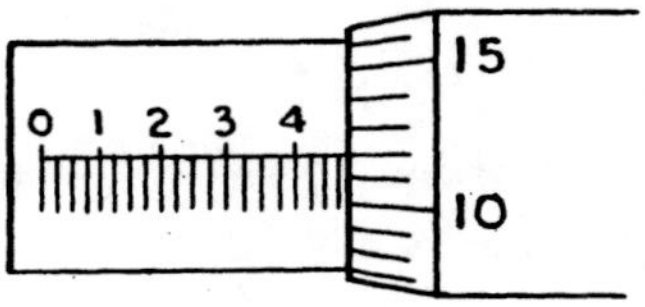

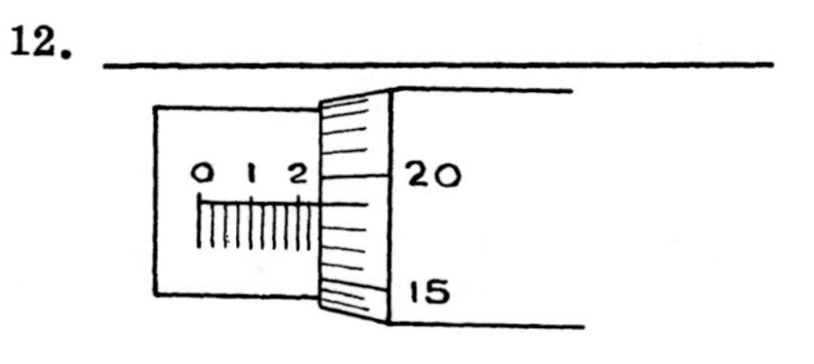

13. ____________________ 14. ____________________

Student's Name________________________ Student Number____________

15. In the micrometer reading shown in question 13, if the thimble was turned clockwise one complete turn the reading would be—

    a. .452
    b. .462
    c. .512

16.

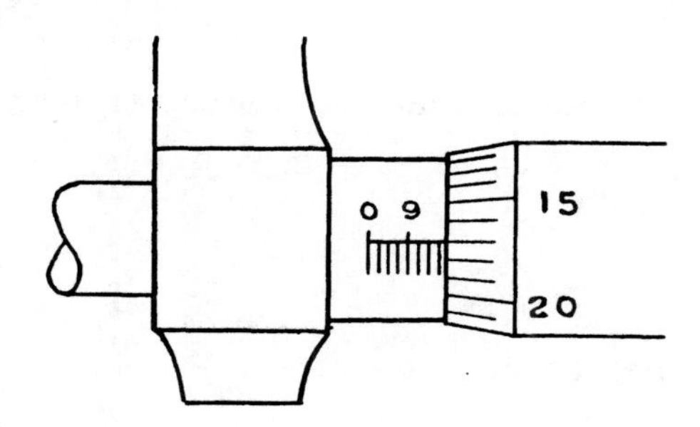

The correct reading of this Inside Micrometer Caliper is—

    a. .842
    b. .822
    c. .817

17.

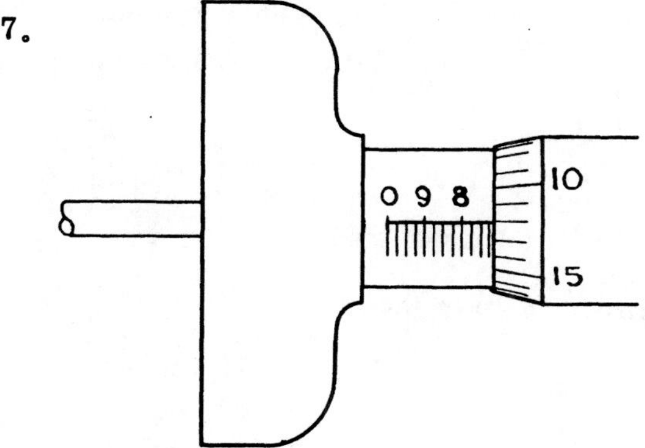

The correct reading of this Micrometer Depth Gage is—

    a. .712
    b. .718
    c. .825

Directions: Study the Vernier Micrometer readings then write the correct reading on the line provided under each illustration.

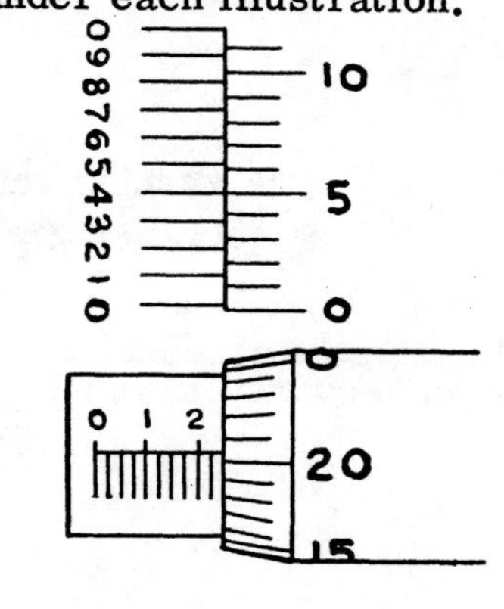

18. ________________________

19. ________________________

(OVER)

CUT OFF HERE

20. To obtain a reading of 6 millimeters on a metric micrometer, starting from the closed position or a zero reading, the thimble must be turned in a counter-clockwise direction—

    a. 6 complete revolutions
    b. 12 complete revolutions
    c. 5 complete revolutions

21. Each graduation or line on the thimble of a metric micrometer equals—

    a. .01 millimeter
    b. .1 millimeter
    c. .5 millimeter

22. The correct reading of this vernier caliper is—

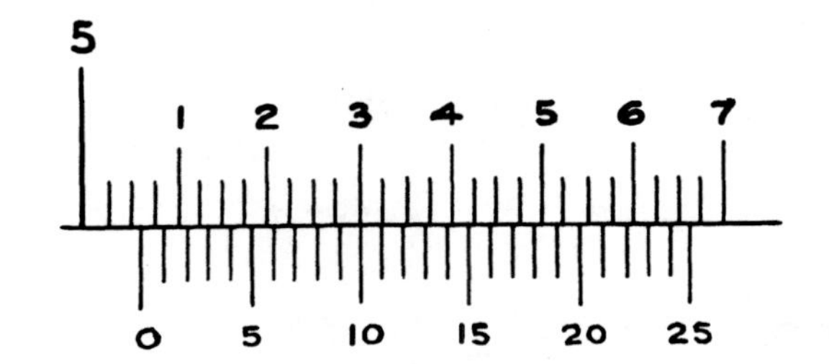

    a. 5.300 inches
    b. 5.050 inches
    c. 5.060 inches

23. The correct reading of this vernier caliper is—

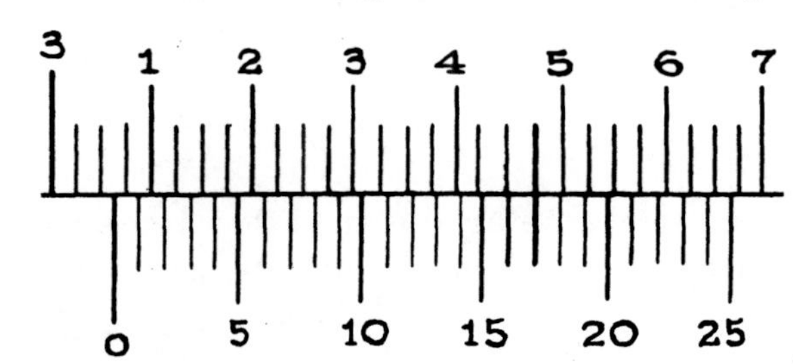

    a. 3.475 inches
    b. 3.067 inches
    c. 3.050 inches

24. A D A vernier caliper cannot be used to measure parts where an accuracy greater than .001 is required.

25. A_______________ _______________gage is a tool used in laying out work to an accuracy of .001.

Student's Name____________________ Student Number__________

26. The correct reading of this Vernier Bevel Protractor is—

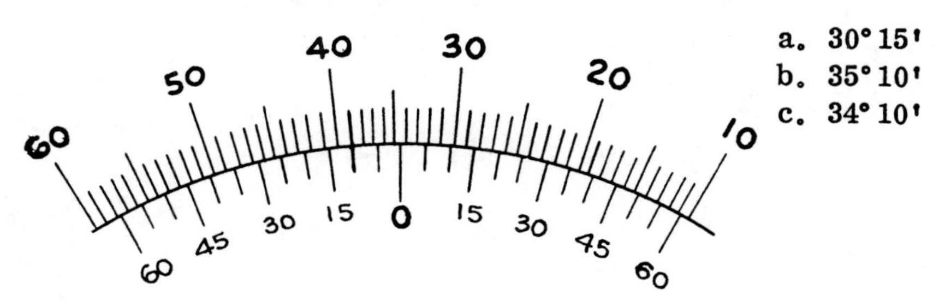

a. 30° 15'
b. 35° 10'
c. 34° 10'

27. The correct reading of this Vernier Bevel Protractor is—

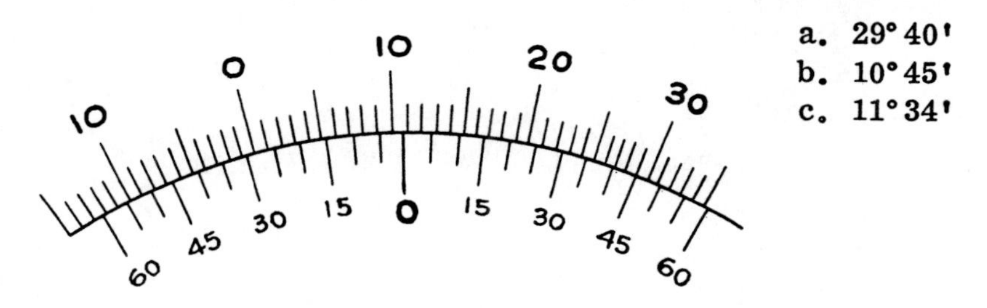

a. 29° 40'
b. 10° 45'
c. 11° 34'

28. A D A clockwise movement of the pointer hand, when checking work with a Dial Indicator, indicates an increase in part size.

29. A D Any variation in part size less than .001 cannot be determined accurately from a Dial Indicator.

30. A D Dial indicators are frequently used, not only for checking machined parts for size variation but also in the setup of such parts before machining.

(END OF EXAMINATION)

CUT OFF HERE

SECTION 2 — Study Guide for MACHINE SHOP OPERATIONS AND SETUPS

TEXTBOOK CHAPTER COVERED

CHAPTERS 3 AND 4 — Bench Tools, Layout Tools, and Power Saws

CONTENTS OF SECTION 2

Assignment 1 and Progress Quiz 6 — Study textbook pages 50 to 75, the machinists' use of hand tools including hammers, files, chisels, wrenches, taps and reamers.

Assignment 2 and Progress Quiz 7 — Study textbook pages 75 to 89, the use of layout tools including the steel rule, scriber, dividers, trammels, prick and center punches, the combination set, and the surface gage.

Assignment 3 and Progress Quiz 8 — Study textbook pages 89 to 105 covering power hack saws, vertical and horizontal band saw machines, the setup and operation of these power saws.

EXAMINATION 2 — Based on pages 50 through 105 in the Assignments for SECTION 2 of this STUDY GUIDE.

WHAT THE STUDY OF SECTION 2 WILL DO FOR YOU

The assignments in this section of the Study Guide introduce you to the many hand tools that the machinist must employ daily in the performance of his job. Manual skill in the use of such hand tools can be developed only through constant use and practice. The study of this section will alert you to the many important points which must be observed and learned to acquire this manual skill rapidly when the occasion arises. Skillful and efficient use of these tools involves a knowledge of what determines the proper selection of a tool for a specific job, how the tool must be held when in use, what safety precautions must be observed, and, above all, how to keep tools clean and sharp.

The final assignment of this section will provide you with the understanding necessary to become skillful in the operation and set up of power hack saws, and vertical and horizontal type band saw machines.

Now begin your study of Assignment 1 on the next page of this Study Guide.

MACHINE SHOP OPERATIONS AND SETUPS

ASSIGNMENT 1 and PROGRESS QUIZ 6

SUBJECT Bench Tools Used in the Shop (Hammers, Files, Chisels, Wrenches, Taps and Reamers) Including Power Driven Saws

TEXTBOOK PAGES TO STUDY 50 to 75

"YOUR AIM" for this assignment will be to acquire the basic knowledge the machinist must have if he is to develop manual skill in the use of hand tools so important to his job in the shop.

SOME SUGGESTIONS

In spite of all the power tools available in the shop, many jobs performed by shop men still require the skillful and efficient use of hand tools, so do not take the study of this assignment lightly. As you study, try to remember the important points which apply to each tool. For example, what precaution must be observed when clamping machined parts in a vise? How do I determine the proper file for a job? When straight filing, should pressure be applied on both forward and return strokes? How do I grind a worn chisel to restore sharpness? Watch for points such as these as you study each tool. Observe, at all times, the safety precautions which may apply to certain tools or operations. So you understand fully any text discussion, be sure to refer to all illustrations as well as any tables in the Appendix as such references are made. These references must be made immediately so you can benefit fully from them.

When you complete the study of this assignment, answer the questions in Progress Quiz 6.

PROGRESS QUIZ 6

THE PURPOSE OF THIS QUIZ: To see how well you understand the correct use and care of all hand tools discussed in this assignment

1. A D False jaws are used on bench vises as a covering to protect the regular jaws of the vise from nicking or other damage. (See textbook, pages 52-53.)

2. A D When filing soft metals such as brass or bronze, use a coarse or rough cut file in preference to a second cut file. (Page 56)

3. A D In conventional or "straight" filing, pressure must be applied only on the forward stroke of the file to prevent dulling the teeth. (Page 57)

4. A D For a fine finish when draw filing, a double-cut file is preferred to a single cut. (Page 58)

(OVER)

5. After they have been used, files should be cleaned with a stiff brush or a ______________ ______________. (Page 58)

6. When chipping metal, you must wear________________ ________________. (Page 60)

7. A tool used in threading a hole is called a______________. (Page 64)

8. A D The correct tap drill size for a $\frac{1}{4}$ - 20 tapped hole is a number 7 drill. (Page 67)

9. A tool used in cutting an external thread is called a__________. (Page 69)

10. A tool used to finish drilled holes accurately to size is called a ________________. (Page 71)

11. The number of teeth per inch on a hack-saw blade is called the__________ of the blade. (Page 72)

12. A D Hack-saw blades should be mounted so that the teeth point away from the saw handle. (Page 73)

13. A D When hack-sawing, you should apply work pressure only on the forward stroke, to prevent dulling the blade. (Page 74)

(END OF QUIZ)

SUGGESTION: Now check your answers to these quiz questions by using the Answer Key provided in the back of this Study Guide. Review the study of any points missed by referring to page numbers given after each question. When you complete this review, begin the study of Assignment 2.

MACHINE SHOP OPERATIONS AND SETUPS

ASSIGNMENT 2 and PROGRESS QUIZ 7

SUBJECT Layout Work and Use of Layout Tools

TEXTBOOK PAGES TO STUDY 75 to 105

"YOUR AIM" for this assignment will be to learn what the operation of "laying out work" has reference to, how it is performed, what tools are used as well as how they are used.

SOME SUGGESTIONS

As you study this assignment, pay particular attention to each illustration, showing the use of layout tools in scribing guide lines on work to indicate the areas to be machined. A complete understanding of layout work so important to the shop man, cannot be achieved without a careful check of the illustrations. Note also that all of the layout tools shown must be used in conjunction with a steel rule. From your study of Chapter 2 in the textbook, you should remember that the laying out of work in this assignment can be accurate only to $\frac{1}{64}$ inch. When a greater accuracy than this is required as in die layouts, the vernier height gage discussed on pages 37 and 38 of your textbook will have to be used. Keep this point in mind as you study and learn the correct use of each layout tool in this assignment.

PROGRESS QUIZ 7

THE PURPOSE OF THIS QUIZ: To test your understanding of the use of layout tools in laying out work for machining.

1. To lay out work, is to scribe lines on metal, to show the areas to be ____________________. (See textbook, page 75)

2. Layout work must be performed on a ________________ ______________ to insure an accurate layout. (Page 76)

3. The tool used to draw a straight line parallel to the edge of a rule is called a__________________. (Page 78)

4. Center punches are used in layout work to locate and mark the centers of __________________. (Page 79)

5. The tool used to scribe a circle in laying out a hole is called a____________. (Page 79)

(OVER)

6. The tool used in laying out center lines on round shafts is called a ____________________rule. (Page 81)

7. A D The center head can be used only to layout and locate centers on the ends of round shafts. (Page 82)

8. A D The hermaphrodite caliper can be used to scribe lines parallel to the edges of a part. (Page 83)

9. Work which must be held in a vertical position for laying out should be clamped to an________________ ________________. (Page 84)

10. The correct height adjustment of a surface gage must be obtained from the rule of a__________________________ ________________. (Page 86)

(END OF QUIZ)

SUGGESTION: Now check your answers to these quiz questions by using the Answer Key provided in the back of this Study Guide. Review any points missed by referring to page numbers given after each question. After this review begin the study of Assignment 3.

MACHINE SHOP OPERATIONS AND SETUPS

ASSIGNMENT 3 AND PROGRESS QUIZ 8

SUBJECT Power Saws

TEXTBOOK PAGES TO STUDY 75 to 105

"YOUR AIM" in this assignment will be to learn about power saws, how they operate, and the type of work they perform.

SOME SUGGESTIONS

The power saw as first designed was rather limited in the work it could perform and because of this was primarily used to cut off stock to prescribed lengths. This stock was then machined to different shapes and forms on other machine tools. As you read through this assignment, note that power saws are now capable of performing production work normally done by many different types of machine tools. A careful study of both the textbook material and all illustrations will enable you to learn about the different types of these machine tools available, the type of work they are capable of performing, what to consider in the selection of a saw blade for specific work, and how to weld these saw blades.

PROGRESS QUIZ 8

THE PURPOSE OF THIS QUIZ: To test your understanding of the setup and operation of the different types of band saws and their construction features

1. A power saw with a reciprocating cutting motion is called a power________ saw. (Page 90)

2. A power saw which employs a continuous looped blade which is driven by two wheels or pulleys is called a____________machine. (Page 91)

3. Contoured or curved cuts can only be performed on______________type band saw machines. (Page 91)

4. To obtain specific lengths of cut on a power feed saw table, the table is provided with____________. (Page 94)

5. When welding saw blades, the overlapping of the saw blade ends can be prevented by proper______________adjustment. (Page 95)

6. To eliminate the brittleness which occurs due to welding, the weld must be________________. (Page 95)

7. The number of teeth per inch in a band saw blade is referred to as ________________. (Page 96)

8. When band sawing a part to a contoured layout line and a filed finish is required, the saw cut should be made____________outside of the layout line. (Page 99)

9. When laying out the contours of a part for sawing, it is advisable to ______________ ______________the layout line, as the layout line may rub off. (Page 101)

10. The wearing of______________ ______________is always advisable when operating a band saw. (Page 104)

(END OF QUIZ)

Now answer all questions in Examination 2.

CUT OFF HERE

MACHINE SHOP OPERATIONS AND SETUPS

Examination 2

TOPIC: Bench Tools, Layout Tools, Power Saws

Based on pages 50 to 105 of the textbook

Student's Name ______________________ Student Number __________

Street ______________ City ______________ State __________ Zip Code ______

1. A file which has parallel rows of teeth that cross one another is called a __________ __________ file.

2. The proper cut of file to use when filing soft metal such as brass is the—

   a. smooth-cut file
   b. second-cut file
   c. coarse- or rough-cut file

3. When "rough" filing, apply pressure only on the __________ stroke of the file, to prevent dulling the teeth.

4.  The method of filing shown at the left is known as __________ filing. When filing in this manner, use a __________ cut file.

Directions: Identify each chisel illustrated by writing the correct name on the line provided under each illustration.

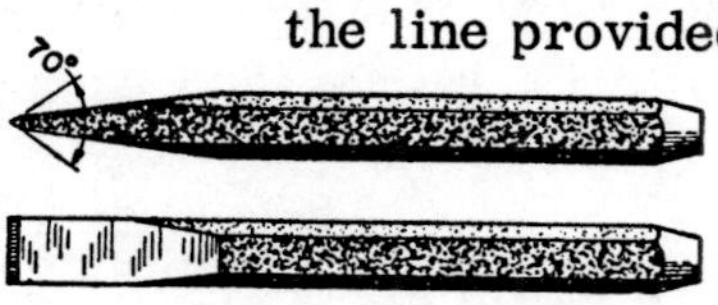

70°

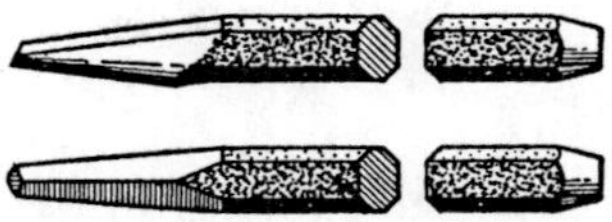

5. ______________________ 6. ______________________

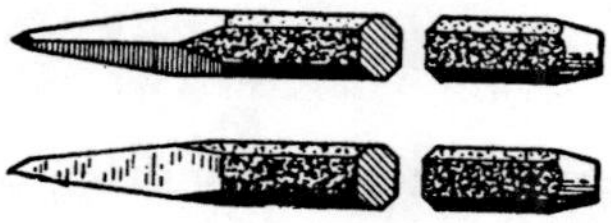

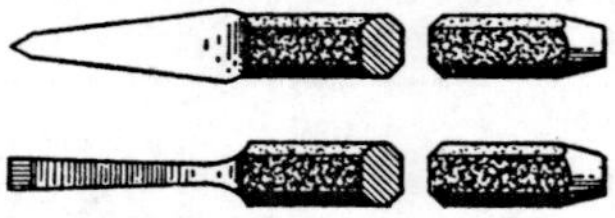

7. ______________________ 8. ______________________

F4-4

(OVER)

9. Flat chisels, to be used for chipping cast iron, should be ground so that the facets forming the cutting edge are at equal—

   a. 70 degree angles
   b. 40 degree angles
   c. 30 degree angles

10. A D When tightening a nut with any wrench, it is safer to apply a pulling rather than a pushing force to the wrench.

11. A wrench which has one solid jaw and one movable jaw is called an ____________________ wrench.

12. When using a wrench of the type described in question 11, always apply force in the direction of the ________________ jaw.

13. A D Set-screw wrenches are more commonly referred to as Allen wrenches, in the shop.

14. A hole which does not go entirely through a workpiece is referred to in the shop as a ________________ hole.

15. A D The size drill to use for a specified tap size is not important as long as it is smaller than the tap diameter.

16. To prevent injury, always wear ______________ ________________ when working with a punch or chisel to remove a broken tap from a hole.

17. A tool used in cutting an external thread is called a—

   a. threading tool
   b. tap
   c. die

18. A tool used to finish drilled holes accurately to size is called a ________________.

19. The term which refers to the number of teeth per inch in a hack-saw blade is ________________.

20. The ____________ of properly mounted hand hack-saw blades point away from the handle of the saw.

Student's Name______________________________ Student Number____________

21. What saw blade pitch is recommended as most suitable when sawing sheet metal less than 18 gage thickness?

    a. 24 pitch
    b. 18 pitch
    c. 32 pitch
    d. 14 pitch

22. When you use a hand hack-saw, apply pressure—

    a. only on the forward stroke
    b. on both strokes
    c. only on the return stroke

23. This is an illustration of a____________ ____________used in layout work to make small indentations at a point where two layout lines____________.

30°

24. This is an illustration of a____________ ____________. It is used in layout work to mark the centers of holes which are to be ____________out later.

90°

25.

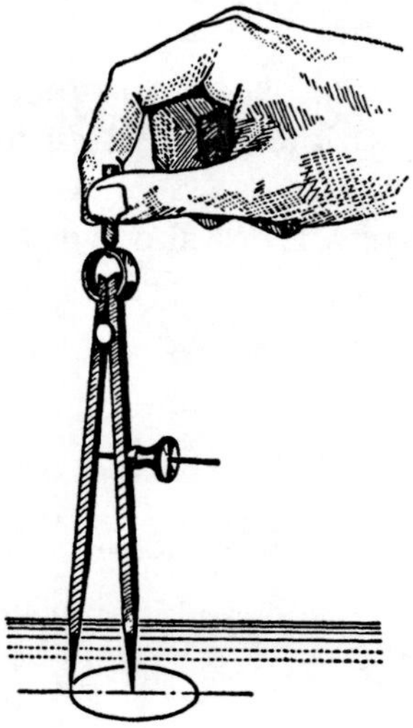

This illustration shows the correct use of____________for____________ circles in hole layout.

(OVER)

CUT OFF HERE

26. This illustrates the proper use of a ________________ ________ when laying out or scribing lines ________________ to a shouldered surface or recess.

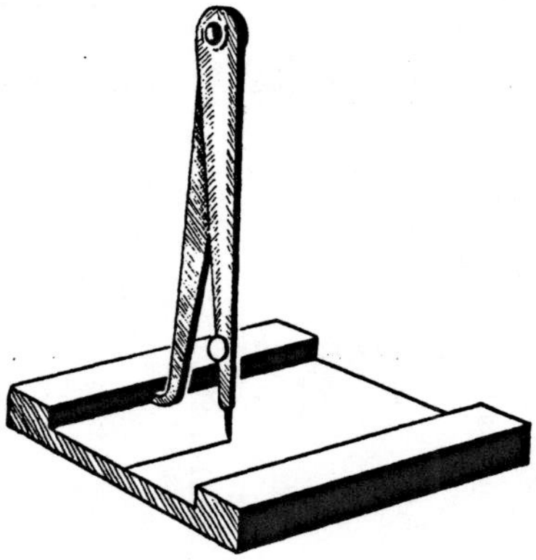

27. The cutting action of a power hack saw blade is described as—

    a. an oscillating motion
    b. a reciprocating motion
    c. continuous circular motion

28. A power saw having a continuous looped blade driven by two pulleys or wheels is referred to as a—

    a. band machine
    b. jig saw
    c. hack saw
    d. swing saw

29. Band sawing parts to some irregular shape or form following a layout line is called________________sawing.

30. Starting holes which are used when band sawing holes in piece parts, should always be drilled________________to the layout line for the hole.

31. Band sawing a multiple number of piece parts from sheet stock is referred to as________________sawing.

(END OF EXAMINATION)

SECTION 3 — Study Guide for MACHINE SHOP OPERATIONS AND SETUPS

TEXTBOOK CHAPTER TO STUDY

CHAPTER 5 — Drill Presses: Types, Setups and Operations

CONTENTS OF SECTION 3

Assignment 1 and Progress Quiz 9 —	Study textbook pages 106 to 121, various types of drill presses and their construction, drills, and drilling fundamentals and the basic cutting action of a twist drill.
Assignment 2 and Progress Quiz 10 —	Study textbook pages 122 to 137, speeds, feeds and coolants for drilling, drill grinding, analyzing drilling difficulties and cutting tools other than drills used on drill presses.
Assignment 3 and Progress Quiz 11 —	Study pages 137 to 150, work and tool holding devices, drill press operations and setups, safety precautions.

MEMORY JOGGER 1

EXAMINATION 3 — Based on pages 106 to 150 in the Assignments for Section 3 of this STUDY GUIDE

WHAT THE STUDY OF SECTION 3 WILL DO FOR YOU

The study of this section will introduce you to the most frequently used machine tool in the shop. In this section you will learn exactly how this machine functions and what operations are performed on it other than drilling. A complete

(OVER)

understanding of drill press work requires knowledge of all cutting tools used, the correct selection of speeds and feeds, the correct way to grind a drill, the various work and cutter-holding devices used, and the safety precautions to be observed when using this machine. This knowledge will help you acquire skill in the operation of the drill press.

Now begin your study of Assignment 1 on the next page of this Study Guide.

MACHINE SHOP OPERATIONS AND SETUPS

ASSIGNMENT 1 AND PROGRESS QUIZ 9

SUBJECT Basic Types of Drill Presses and Construction, Drills and Drilling Fundamentals, and the Cutting Action of a Twist Drill

TEXTBOOK PAGES TO STUDY 106 to 121

"YOUR AIM" for this assignment will be to learn the operating characteristics and the differences in construction of the four basic types of drill presses found in the shop, the functional elements of a twist drill, and finally how these elements contribute to the cutting action of a twist drill.

SOME SUGGESTIONS

The drill press, as you will find out in the study of this assignment, is not a difficult machine to operate because it has so few controls and movements. Efficient operation or skill is more dependent on the knowledge you will acquire in the remaining assignments of this section, than on knowledge of the drill press itself. To develop a working knowledge of this machine, study the illustrations of each type of drill press. All of the control points as well as all the important parts of the machine are clearly labeled. The discussion on drills and drilling fundamentals which begins on page 112 of your textbook is very important, for the twist drill is the primary cutting tool used on a drill press. A complete understanding of this cutting tool and how it must be sharpened is one of the factors which will contribute to your ability in the skillful operation of a drill press. Be sure to refer to Table IV in the Appendix as mentioned on page 115 of the textbook, for more detailed information concerning drill point geometry.

When you complete the study of this assignment, answer all questions in Progress Quiz 9.

PROGRESS QUIZ 9

THE PURPOSE OF THIS QUIZ: To test your understanding of the twist drill and the construction of the four basic drill press types

1. A D Skill in drill press work requires only a working knowledge of the drill press itself. (See textbook, page 106)

2. A D The sensitive drill press is the only drill press not equipped with power feeds. (Pages 106-107)

(OVER)

3. A D Radial and Multi-Spindle drill presses are provided with both manual and power feeds. (Page 111)

4. A D To reduce the feeding pressure needed for drilling larger holes, it is good practice first, to drill a small lead hole. (Page 113)

5. The_______________ _____________of a twist drill are called lips. (Pages 113-114)

6. It is necessary when grinding drills to provide the lips with______________ so they can cut. (Page 114)

7. The two helical grooves which extend the full length of the drill body are called________________. (Page 115)

8. To check the diameter of a drill with a micrometer, the measurement must be taken across both__________________of the drill. (Page 115)

9. A D The point angle of a drill must be varied by grinding to suit the material to be drilled. (Page 115)

10. Sizes of twist drills are classified by________________, fractions, and ________________. (Page 117)

11. A D The exact size of a twist drill can be measured only with a drill gage. (Page 118)

12. A D A factor which reduces the efficiency of a twist drill as a cutting tool is its web or chisel edge. (Page 119)

13. A D The cutting action of a twist drill can be described as a shearing action. (Page 120)

(END OF QUIZ)

SUGGESTION: Check your answers to these quiz questions by using the Answer Key provided in the back of this Study Guide. Review the study of any points missed by referring to page numbers given after each question. When you complete this review begin the study of Assignment 2.

MACHINE SHOP OPERATIONS AND SETUPS

ASSIGNMENT 2 AND PROGRESS QUIZ 10

SUBJECT Selection of Speeds, Feeds, and Coolants for Drilling, Proper Drill Grinding, All Cutting Tools Used on Drill Presses

TEXTBOOK PAGES TO STUDY 122 to 137

"YOUR AIM" for this assignment will be to learn what factors must be considered in the correct selection of speeds, feeds, and coolants for drill press operations and why this is so important to the skillful operation of this machine tool. You will also learn what cutting tools other than twist drills are used on the drill press.

SOME SUGGESTIONS

To develop skill or be able to operate a drill press efficiently, you must have a complete knowledge and understanding of what speeds and feeds must be used. As you study this assignment, try to remember what factors determine the speed and feed to use and how the surface speed and revolutions per minute of a drill can be calculated if necessary, or found by reference to tables in the Appendix of the textbook. Proper use of these tables is important so be sure to refer to them as you study pages 122 and 123 of the textbook. On page 128 of the textbook is a handy check list of drilling difficulties you may encounter on the job and what must be done to correct the problem. The ability to analyze a drilling problem and knowledge of what must be done to correct the problem are of prime importance in the operation of a drill press. Of equal importance is the discussion on pages 129 to 137 of the textbook, of types of cutting tools other than drills used on the drill press.

To see how well you can recall the important points of this assignment, answer all questions in Progress Quiz 10.

PROGRESS QUIZ 10

THE PURPOSE OF THIS QUIZ: To test your knowledge of the factors important to the operation of a drill press

1. A D The type of drill being used and the material to be drilled are the two factors which determine the proper cutting speed to use. (See textbook, page 122.)

2. A D The diameter of the drill is a factor which determines the R.P.M. needed to obtain a specific cutting speed. (Page 123)

(OVER)

3. Kerosene is a good cutting fluid to use when drilling_______________.
(Page 124)

4. A D A cutting fluid does not have to be used when drilling or reaming cast iron. (Page 124)

5. When you are grinding a drill, it is important that the_____________and _____________of the lips are ground equal. (Page 124)

6. A cutting tool used to provide a recess at the top of a drilled hole for a flathead machine screw is called a_________________. (Page 129)

7. Drills used in drilling pre-cast or cored holes in castings are called ____________drills. (Page 131)

8. A tool used to finish drilled holes to exact size is called a____________.
(Page 131)

9. A D Rose reamers rather than fluted reamers are designed for taking heavier cuts and are not classified as finishing reamers. (Page 133)

10. A D Shell reamers are not provided with shanks and must be fitted with special arbors for use. (Page 135)

11. A D Hand reaming produces more accurately sized and finished holes than machine reaming. (Page 136)

12. A D A spiral fluted reamer has less tendency to chatter than a straight fluted reamer. (Page 136)

13. A D The same speed should be used in reaming a hole as is used for drilling. (Page 137)

(END OF QUIZ)

SUGGESTION: Check your answers to these quiz questions by using the Answer Key provided in the back of this Study Guide. Review the study of any points missed by referring to page numbers given after each question. When you complete this review, begin the study of Assignment 3.

MACHINE SHOP OPERATIONS AND SETUPS

ASSIGNMENT 3 AND PROGRESS QUIZ 11

SUBJECT Work- and Cutter-Holding Devices, Drill Press Setups and Operations, Safe Practices To Follow

TEXTBOOK PAGES TO STUDY 137 to 150

"YOUR AIM" in the study of this assignment will be to complete your knowledge of the drill press. You will read and learn what work and cutter-holding devices are used and how to make safe and proper setups for reaming tapping, drilling, and spot-facing operations.

SOME SUGGESTIONS

As you study this assignment, keep in mind that all straight-shanked cutting tools used in drill press work must be held in a key type drill chuck. All taper-shanked tools can be mounted directly into the spindle of the drill press or into tapered sleeves or sockets which match the tapered hole of the spindle. For a better understanding of the many work-holding devices shown in Fig. 37 on page 141 of the textbook, refer to Figures 39 and 40 on pages 144 and 145 so you can see exactly how these devices are used in making drill press setups.

When you complete the study of what safety precautions to observe in drill press work, answer all questions in Progress Quiz 11.

PROGRESS QUIZ 11

THE PURPOSE OF THIS QUIZ: To see how well you understand the setup of a drill press to perform specific operations

1. A D All cutting tools used on a drill press must be mounted in a drill chuck for use. (See textbook, page 138.)

2. A D To provide a more positive drive, all taper-shanked tools are equipped with a tang to prevent them from slipping under a cutting load. (Page 138)

3. A D It is not possible to tap a hole on a sensitive drill press without the aid of a tapping attachment. (Page 139)

4. Cylindrical stock should always be clamped in a________ ______________ for drilling. (Page 142)

(OVER)

5. A D Precise drilling of two or more holes can be accomplished only by laying out each hole unless a drill jig is used. (Pages 145-146)

6. The operation of providing a recess at the top of a drilled hole for a flat-head machine screw is called____________________. (Page 146)

7. The operation of providing a recess for the head of a fillister-head screw is called____________________. (Page 147)

8. A D Reamed holes must be drilled $\frac{1}{64}$ less than the reamed size specified on the drawing. (Page 148)

9. A D Holes should be reamed at the same speed that was used in drilling them. (Page 148)

10. When performing any drilling operations, the operator should wear ____________________. (Page 150)

(END OF QUIZ)

SUGGESTION: Check answers by using the Answer Key in the back of this Study Guide. Review the study of any points missed by referring to page numbers given after each question, then proceed to answer the questions in the Memory Jogger on the next page of this Study Guide.

MACHINE SHOP OPERATIONS AND SETUPS

MEMORY JOGGER 1

THE PURPOSE OF THIS QUIZ: To determine how well you can recall the important points you have studied in Chapters 1 through 5 in your textbook

1. A D In measuring the diameter of round stock, it is best to use the slide caliper rule. (Page 15)

2. A D The accuracy of measurements made with inside or outside calipers depends entirely on a keen sense of touch. (Page 20)

3. The decimal equivalent of the fraction $\frac{5}{8}$ is_____________. (Page 24)

4. The decimal equivalent of the fraction $\frac{1}{4}$ is_____________. (Page 25)

5. Each line on the barrel of a micrometer represents_____________of an inch. (Page 27)

6. A D The thimble reading of a plain micrometer is obtained by taking the value of the line on the thimble which matches or aligns with the index line on the barrel. (Page 29)

7. When an accuracy to four decimal places is required in measurement, a _____________________micrometer must be used. (Pages 36-37)

8. A D Each line or division on the vernier scale of a vernier caliper equals .0001 inch. (Page 38)

9. A D The vernier height gage is primarily used in layout work and is read in the same manner as a vernier caliper. (Page 43)

10. An instrument used in measuring angles to an accuracy of five minutes is called a__________________ __________________ ____________________. (Page 44)

11. A D Dial indicators cannot be used to determine the actual size of a part. They can be used only to detect a variation in size. (Page 46)

12. A D When straight filing, you should apply pressure on both the forward and return strokes. (Page 57)

13. The tool used in threading a hole is called a_____________. (Page 64)

(OVER)

14. A tool used in cutting external threads on round stock is called a________.
(Page 69)

15. A D The kind of metal to be sawed is one of the factors which determines the correct pitch blade to use in a hand hack saw.
(Page 72)

16. Scribing lines on metal to indicate the areas to be machined is called ________________work. (Page 75)

17. The tool used to scribe small circles in hole layout is called a__________.
(Page 80)

18. Cutting stock to prescribed length is typical of the work performed by ____________________band saws. (Page 91)

19. Piece parts having an irregular shape or form are frequently cut to size on the________________band saw. (Pages 91-92)

20. To obtain an angle cut on a vertical band machine, the_____________must be tilted. (Page 94)

21. When the blade of a vertical band machine is being welded, an overlapping of the saw blade ends indicates an incorrect adjustment of the jaw pressure selector for________________. (Page 95)

22. To eliminate brittleness from a saw blade weld, the weld must be __________________. (Page 95)

23. The_____________of a saw blade refers to the number of teeth per inch.
(Page 96)

24. For an accurate adjustment of the vertical saw band________________, a feeler gage must be used. (Page 98)

25. As a matter of safe practice when operating a band saw, wear___________ ______________at all times. (Page 104)

26. To prevent nicks and cuts when handling saw band blades, wear__________.
(Page 105)

(END OF QUIZ)

SUGGESTION: Now check your answers by using the Answer Key in the back of this Study Guide. Review all points missed; then answer all questions in Examination 3.

MACHINE SHOP OPERATIONS AND SETUPS

TOPIC: Drill Press Setups and Operations

Examination 3

Based on pages 106 to 150 of the textbook

Student's Name____________________ Student Number____________

Street____________ City____________ State__________ Zip Code________

1. Because they are essentially high speed machines, sensitive drill presses should not be used for drilling hole diameters greater than-

   a. 1 inch
   b. 3/4 inch
   c. 7/8 inch
   d. 1/2 inch

2. The only drill press not equipped with a power feed for the spindle is the-

   a. vertical drill press
   b. radial drill press
   c. multi-spindle drill press
   d. sensitive drill press

3. A distinguishing feature of______________drill presses is that they have a moveable spindle head which can be positioned at varying distances from the drill press column.

4. The purpose of the____________on a taper-shanked drill is to prevent drill slippage and to assure a more positive drive.

5. The point and clearance angles of drills must be varied by grinding dependent upon the-

   a. drill diameter
   b. material drilled
   c. rotational drill speed used
   d. hole depth to be drilled

6. Under the numbering classification of twist drill diameters, which of the following sizes would be considered as largest?-

   a. No. 25
   b. No. 18
   c. No. 30
   d. No. 44

7. A twist drill gage is used to measure-

   a. drill length
   b. drill clearance angles
   c. drill diameter
   d. point angles of drills

F4-4

(OVER)

CUT OFF HERE

8. To drill free machining stainless steel at a cutting speed of 40 ft. per minute with a 5/8 diameter drill, the drill must rotate at an r.p.m. of-

   a. 105   c. 244
   b. 306   d. 350

9. This illustration shows the correct use of a__________ __________gage.

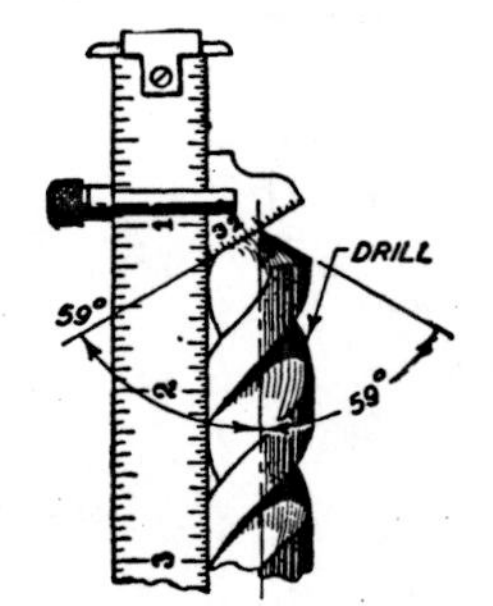

It is used for checking both the__________ and the__________of the cutting lips of a drill.

10. It is not necessary to use a cutting fluid or lubricant when drilling-

   a. cast iron   c. copper
   b. aluminum   d. steel

11. The tool shown in this illustration is called a-

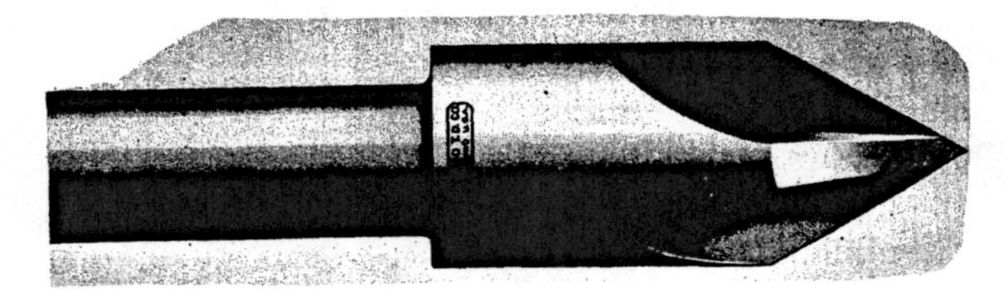

   a. spotfacer
   b. countersink
   c. counterbore
   d. core drill

12. To provide a close tolerance and a fine finish in a drilled hole, the hole must be__________________.

13. The cutting tool illustrated above is called a-

   a. shell reamer
   b. hand reamer
   c. taper-shanked solid reamer
   d. rose reamer

Student's Name________________________Student Number________________

14. A D To ream a hole which has just been drilled, the machine operator must remember to reduce the speed of the machine.

15. All straight-shanked twist drills must be mounted or held in a-

a. drill socket
b. tapered sleeve
c. key-type drill chuck

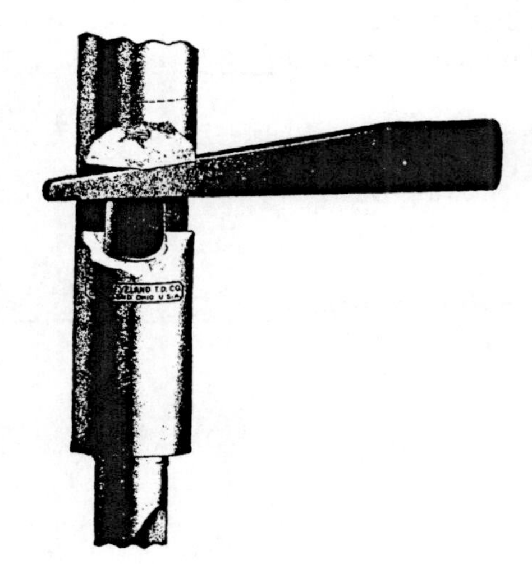

16. This illustration shows the correct use of a____________pin. It must be used to remove all____________shanked tools from a drill press spindle.

17. Power tapping of holes on____________ type drill presses requires the use of a tapping attachment.

18. The work-holding device shown here is called a________ ____________.

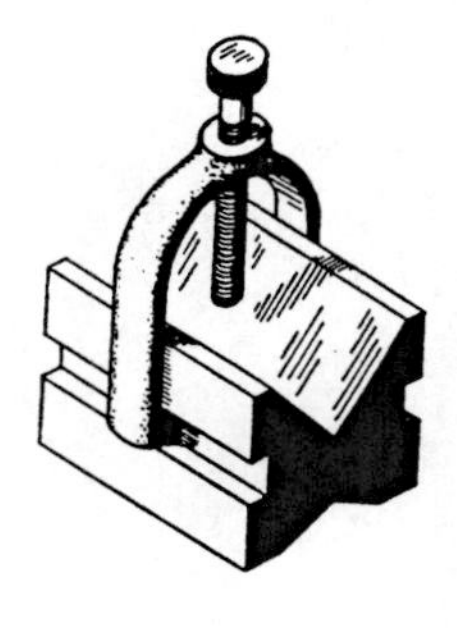

This device is used primarily for holding work which is____________________in shape.

19. A D When drilling work held in a vise, it is unnecessary to clamp the vise to the drill press table as the vise can be held by hand.

(OVER)

CUT OFF HERE

20. The operation of providing a cone-shaped recess at the top of a drilled hole for a flathead machine screw is called______________________.

21. Countersinks being a straight shanked tool are mounted for use in a ______________ ______________.

22. To assure proper alignment of a counterbore with a previously drilled hole, the counterboring tool is provided with a______________.

23. The operation of providing a smooth seat or bearing surface around a previously drilled hole for a washer or nut is called______________.

24. The drill size to be used for a 1/4 inch reamed hole is-

    a. 15/64
    b. 7/32
    c. 17/64
    d. 13/64

25. Broken drills frequently result from using excessive______________and ______________and therefore should be avoided.

(END OF EXAMINATION)

SECTION 4 — Study Guide for MACHINE SHOP OPERATIONS AND SETUPS

TEXTBOOK CHAPTERS COVERED

CHAPTERS 6 AND 7 — Engine Lathe Types, Accessories, Attachments, Cutting Tools, and External Lathe Operations

CONTENTS OF SECTION 4

Assignment 1 and Progress Quiz 12 — Study textbook pages 151 to 179, engine lathe construction, accessories, attachments, , lathe maintenance, and safe operating practices.

Assignment 2 and Progress Quiz 13 — Study textbook pages 180 to 199, high speed and carbide cutting tools for lathe use.

Assignment 3 and Progress Quiz 14 — Study textbook pages 199 to 217 covering basic external lathe operations.

EXAMINATION 4 — Based on textbook pages 151 through 217

WHAT THE STUDY OF SECTION 4 WILL DO FOR YOU

To acquire skill in the operation and setup of an engine lathe, you must have a general knowledge of the machine tool itself. This knowledge consists of knowing the correct names and locations of the engine lathe parts and their function; what controls are provided for selection of speeds and feeds; what attachments, accessories, and cutting tools are used; and how to care for and operate a lathe in a safe manner. These are the things you will learn in addition to how to perform basic external lathe operations such as facing, straight turning of work between centers, shoulder turning, knurling, and cutoff operations.

Now begin your study of Assignment 1.

MACHINE SHOP OPERATIONS AND SETUPS

ASSIGNMENT 1 and PROGRESS QUIZ 12

SUBJECT Engine Lathe Construction, Accessories and Attachments Used

TEXTBOOK PAGES TO STUDY 151 to 179

"YOUR AIM" for this assignment is to obtain a general working knowledge of the engine lathe which includes knowing the names of lathe parts, what accessories and attachments are used in lathe work, and how to care for and maintain a lathe in good operating condition.

SOME SUGGESTIONS

Before you begin reading about the parts of a lathe and lathe construction, you should study the illustrations on page 157 in your textbook. This will acquaint you with the many features provided on an engine lathe and will help you understand the textbook discussion covering the function of the various parts as well as the steps involved in setting up a simple shaft for turning on a lathe.

When you finish this study assignment, answer the questions in Progress Quiz 12.

PROGRESS QUIZ 12

THE PURPOSE OF THIS QUIZ: To see how well you understand the function of the various parts of a lathe and what accessories and attachments are used

1. A D In lathe work, the cutting tool feed is expressed in inches per revolution. (See textbook, page 152.)

2. A D Tool room lathes are so classified because they can produce a greater variety of precision work than standard lathes. (Page 152)

3. The unit of a lathe which houses the lathe spindle and control levers for speed selection is called a ____________________. (Page 159

4. When drilling work on a lathe, mount the drill in the spindle of the ____________________. (Page 160)

(OVER)

5. A D The feed rate of a lathe depends upon the setting of the levers on the gear box. (Page 164)

6. A D Work held in a universal chuck is automatically centered when the chuck is tightened. (Page 165)

7. When mounting any large or heavy chuck on the lathe spindle, it should be supported or held in a_______________ _______________on the lathe bed until it is secure on the spindle. (Page 166)

8. Lathe work which cannot be held in a lathe chuck can be clamped to a ____________________mounted on the headstock spindle. (Page 169)

9. To take up wear on the cross slide or compound rest, the adjustable______ must be tightened. (Page 178)

10. A D Long-sleeved shirts are considered a safety hazard when worn by workers who are operating lathes. (Page 178)

(END OF QUIZ)

SUGGESTION: Now check your answers by using the Answer Key in the back of this Study Guide. Review all points missed, then begin your study of Assignment 2.

MACHINE SHOP OPERATIONS AND SETUPS

ASSIGNMENT 2 and PROGRESS QUIZ 13

SUBJECT Lathe Cutting Tools

TEXTBOOK PAGES TO STUDY 180 to 199

"YOUR AIM" for this assignment is to learn how to select the proper cutting tool for a job, how to grind the tool so it will have the proper clearance and rake angles, and how it should be mounted or held in relation to the work when performing different operations.

SOME SUGGESTIONS

In this assignment, be sure to study carefully Fig. 2 on page 182 of the textbook. The quality of fine surface finish required in work machined on a lathe cannot be obtained unless you thoroughly understand the six basic tool angles and their function. Refer to this and all illustrations carefully as the textbook mentions them, to obtain a complete understanding of these various tool angles and the various shapes of cutting tools used for facing, cutting off, and rough and finish turning operations.

After you complete the study of this assignment, answer the questions in Progress Quiz 13.

PROGRESS QUIZ 13

THE PURPOSE OF THIS QUIZ: To see how well you understand the grinding and mounting of all lathe cutting tools for specific lathe operations

1. The angle on the face or top surface of a tool which slants from the side cutting edge is called a____________ ____________angle. (See textbook, page 183.)

2. The angle on the face of a tool which slants backward from the nose of the tool is called a____________ ____________angle. (Page 183)

3. The angles ground into the sides of lathe tools are called relief or ____________________angles. (Page 184)

4. A D The side cutting edges of a left-hand tool are on the right side of the tool. (Page 188)

(OVER)

5. When facing work in a lathe, it is important that the point of the tool be set on the exact_______________of the workpiece. (Page 191)

6. A round-nose tool ground with no_____________rake can be used as a right- or left-hand tool. (Page 193)

7. A D A rough turning tool should be ground with less clearance than is normally ground on a finishing tool. (Page 193)

8. When taking a finish cut, the lathe should be operated at a____________ feed and____________speed. (Page 194)

9. A cutoff tool is similar to a________________tool except that it is narrower. (Page 194)

10. The most wear-resistant grade of carbide used for cutting tools is the straight________________carbide. (Page 196)

(END OF QUIZ)

SUGGESTION: Check answers by using the Answer Key in the back of the Study Guide. Review points missed if necessary, then begin your study of Assignment 3.

MACHINE SHOP OPERATIONS AND SETUPS

ASSIGNMENT 3 and PROGRESS QUIZ 14

SUBJECT External Lathe Operations: Facing, Straight Turning, Shoulder Turning, Knurling, Cutting Off, Filing, Polishing, Grinding, and Honing

TEXTBOOK PAGES TO STUDY 199 to 217

"YOUR AIM" in this assignment is to learn the proper procedures to follow in setting up an engine lathe to perform the various operations listed above under the heading of "SUBJECT."

SOME SUGGESTIONS

As you study this assignment, note that the setup of a lathe to perform each specific operation discussed is covered step by step. The order in which these steps must be performed is very important so keep this in mind and try to remember the correct procedure to follow in each case. Other points to observe carefully in each setup are the work-holding devices employed and what safety precautions apply when mounting or removing them. Proper mounting of the cutting tool in relation to the work will vary with each setup so note this carefully also as well as the type or form of tool which must be used. To refresh your memory on tool forms, it may help to refer to Fig. 12 on page 189 of the textbook when the mounting of the tool is discussed for each operation. Referring to all illustrations will help you understand the reading material so refer to each one as it is mentioned.

When you finish this study assignment, answer the questions in Progress Quiz 14.

PROGRESS QUIZ 14

THE PURPOSE OF THIS QUIZ: To see how well you understand each step or procedure in setting up a lathe to perform specific operations

1. A D To face work properly on a lathe, the work must be mounted in a universal chuck. (See textbook, pages 199-200.)

2. A D To prevent damaging the ways of a lathe when removing a chuck, a block of wood should be placed across the ways, under the chuck. (Page 200)

3. A D The grooves in the face of a four-jaw chuck serve as a guide to center the stock when adjusting each jaw, so the stock will run true. (Page 200)

(OVER)

4. To check accurately if stock is centered in a four-jaw chuck, a__________ __________________is used. (Page 200)

5. A D When facing work in a lathe, the carriage lock screw must be tightened before the cut is taken. (Page 201)

6. Work which cannot be chucked because of its shape can be mounted on a ____________________for facing. (Page 202)

7. The operation of taking a cut along the_______________of a piece of stock is called straight turning. (Page 207)

8. Before work can be centerdrilled, the_______________and_______________ must be checked for alignment. (Page 208)

9. For a turning operation, the tool bit should be set slightly____________the center of the stock. (Page 208)

10. The diameter of a finish turned shaft must be checked for size with a _______________________. (Page 210)

11. When using a knurling tool, set the working faces of the rolls____________ to the work surface. (Page 212)

12. A D Oil must be used as a lubricant for the work surface when knurling. (Page 213)

13. When cutting off work, the lathe________________must be locked to the lathe bed. (Page 213)

14. A D Filing work in a lathe while frequently done, is not considered good practice. (Page 214)

(END OF QUIZ)

SUGGESTION: Now check your answers by using the Answer Key in the back of this Study Guide. Review all points missed, then answer all questions in Examination 4.

CUT OFF HERE

MACHINE SHOP OPERATIONS AND SETUPS **Examination 4**

TOPIC: Engine Lathe: Construction, Cutting Tools, Accessories, Basic External Lathe Operations

Study pages 151 to 217 and the illustration of the Engine Lathe (A) in your textbook

Student's Name____________________ Student Number____________

Street____________ City__________ State________ Zip Code_______

1. The proper____________rate is obtained on a lathe by the setting of the levers on the quick-change gear box.

2. A D Movement of the lathe carriage can be accomplished only by engaging the power feed lever on the apron.

3. Which one of the lathe parts mentioned is not provided with a power feed?

   a. carriage
   b. compound rest
   c. cross slide

4. A D On lathes equipped with both lead screw and a feed rod, the drive power for turning operations is obtained from the feed rod.

5. Parts having a square or irregular shape to be machined on a lathe can only be mounted in—

   a. universal chucks
   b. independent chucks
   c. collet chucks
   d. cam lock chucks

6. When mounting heavy chucks on lathe spindles, place a____________ ____________across the bed of the machine and under the chuck for a safety device.

7. The preferred chuck for mounting small cylindrical work to be turned is the____________chuck.

8. Tailstock centers which do not revolve with the workpiece are referred to as____________centers.

F4-4

(OVER)

9. Print the letter from the illustration next to the lathe part with which it corresponds.

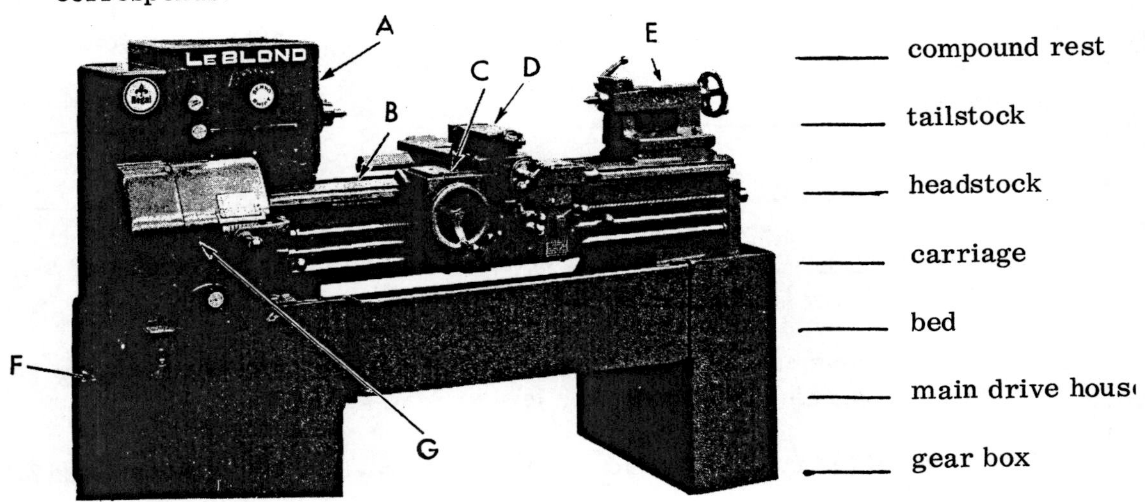

_____ compound rest

_____ tailstock

_____ headstock

_____ carriage

_____ bed

_____ main drive hous

_____ gear box

10. To refinish the points of lathe centers which have become worn, grind them to an included angle of—

    a. 30 degrees
    b. 60 degrees
    c. 90 degrees
    d. 82 degrees

11. A revolving type of center, having a flatted rather than a sharply pointed end, which is used to support the tailstock end of hollow cylindrical stock is referred to as a____________center.

12. Drill chucks when used for drilling operations in a lathe are mounted in the__________________spindle.

13. A device which is fastened to the headstock end of a workpiece when work is turned between centers is called a______________ __________.

14. Workpieces provided with a reamed hole are mounted on a tapered __________________for turning between centers.

Student's Name__________________________ Student Number__________________

15. The lathe tool post (shown in this illustration) which provides rigid support for carbide tools is called—

    a. a quick-change tool post
    b. an open-side tool post
    c. a rocker tool post
    d. a turret-type tool post

16. The dovetail surfaces of the compound rest, the carriage, and the cross slide are provided with adjustable______________to take up wear.

17. A D When lathe attachments are stored after use, they should be coated with a film of oil to prevent rusting.

18. A D Long-sleeved shirts and jewelry are considered to be safety hazards and should not be worn when one is operating a lathe.

19. Write the correct letter which corresponds to each tool bit angle named below.

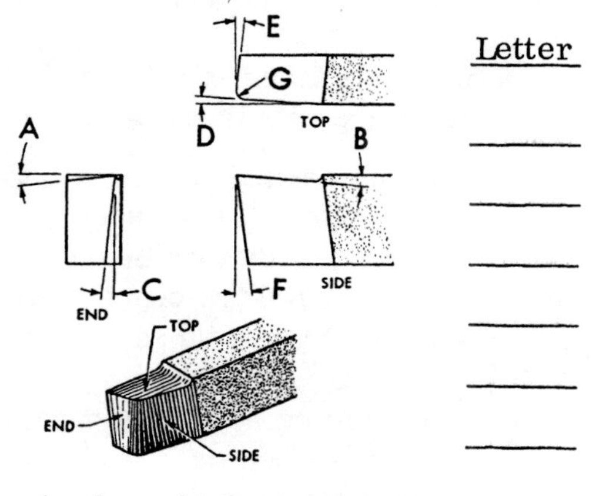

Letter	Angle
______	back rake angle
______	side clearance angle
______	side rake angle
______	front clearance angle
______	end cutting-edge angle
______	side cutting-edge angle

20. Angles which are ground on the top surface or face of a lathe tool bit and which slant downward from the side cutting edge of the tool are called—

    a. positive side rake angles
    b. negative side rake angles
    c. side clearance angles
    d. end cutting-edge angles

(OVER)

CUT OFF HERE

21. Chatter in lathe work which produces a rough surface finish, can be prevented in some cases by regrinding the tool bit to—

a. increase the side cutting-edge angle
b. reduce the back rake angle
c. increase the side clearance angle
d. reduce the side cutting-edge angle

22. Lathe tool bits made of straight tungsten carbide are noted for their ability to resist—

a. cratering
b. galling
c. wear
d. chipping

23. Of the factors listed, which one must be considered in determining the proper grade of carbide tool to use?

a. the type of operation to be done
b. the type of material to be machined
c. the finish requirements of the machined part
d. the speeds to be used in turning the part

24. Extremely accurate centering of work in an independent chuck can only be determined by the use of—

a. a center gage
b. a height gage
c. a dial indicator
d. a surface gage

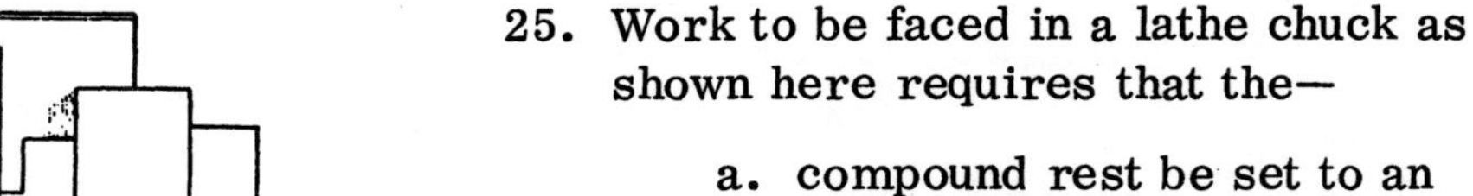

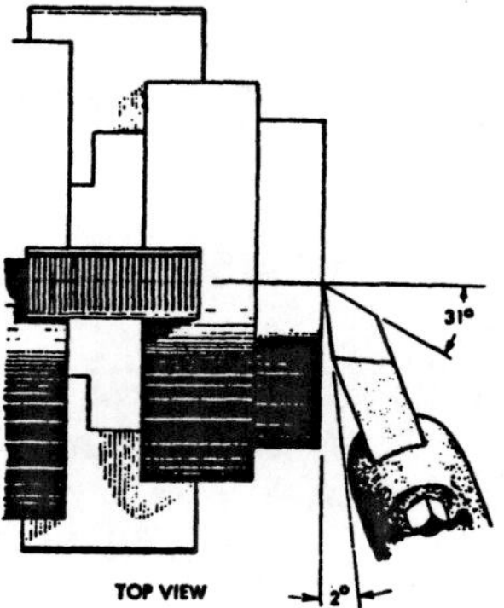

25. Work to be faced in a lathe chuck as shown here requires that the—

a. compound rest be set to an angle of 31 degrees
b. tip of the tool is set exactly on the work centerline
c. tip of tool be set below the work centerline
d. tool be set slightly above the work center

Student's Name________________________ Student Number________________

26. Write the name of the two work-holding devices (labeled A and B in this illustration) which are used when turning or facing work between lathe centers.

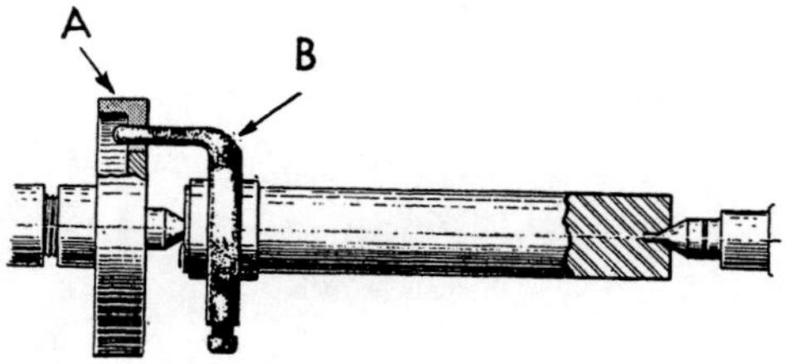

A______________ ______________

B______________ ______________

27. When straight turning a shaft mounted between lathe centers—

    a. set the compound parallel to the work axis
    b. use a lubricant
    c. set the tool bit slightly above the work centerline
    d. set the tool bit below the work centerline

CUT OFF HERE

(END OF EXAMINATION)

SECTION 5 — Study Guide for MACHINE SHOP OPERATIONS AND SETUPS

TEXTBOOK CHAPTERS COVERED

CHAPTERS 7 and 11 — Engine lathe operations of thread cutting, taper turning, drilling, reaming, and boring; production turning as performed on turret lathes and automatic screw machines

CONTENTS OF SECTION 5

Assignment 1 and Progress Quiz 15 — Study textbook pages 217 to 236, thread terms, the operation and setup of engine lathes when cutting unified, acme, and square threads.

Assignment 2 and Progress Quiz 16 — Study textbook pages 236 to 255, the setup and operation of engine lathes for taper turning, drilling, reaming, boring, and internal threading.

Assignment 3 and Progress Quiz 17 — Study textbook pages 368 to 410, covering the types of work and tooling used to perform various operations on turret lathes and automatic screw machines.

EXAMINATION 5 — Based on textbook pages 217 through 255 and Chapter 11 (pages 368 to 410)

WHAT THE STUDY OF SECTION 5 WILL DO FOR YOU

The first two assignments of this section are a continuation of engine lathe operations and will provide you with the knowledge which skilled shop personnel must acquire with respect to thread cutting principles and procedures, tapers and taper turning, drilling, reaming, and boring operations. A close study of this section of the textbook will teach you how to perform these operations efficiently whenever the occasion arises. At this point you are asked to skip to Chapter 11 (starting on page 368) to study the tooling used when operating and setting up turret lathes and automatic screw machines. This was done because of the similarity of these machines to a basic engine lathe (see page 372 of the textbook). The big difference between these types of lathes is that turret lathes are constructed and tooled so that several different operations can be performed at the same time. This naturally increases production and is a great advantage when many similar parts are to be machined. The turret lathe tooling and lathe construction is what makes this possible.

Now begin your study of Assignment 1.

MACHINE SHOP OPERATIONS AND SETUPS

ASSIGNMENT 1 and PROGRESS QUIZ 15

SUBJECT Lathe Setups for Cutting Various Forms of Screw Threads

TEXTBOOK PAGES TO STUDY 217 to 236

"YOUR AIM" for this assignment is to learn how to setup the lathe to perform thread-cutting operations for the Acme, Square, and Unified forms of thread.

SOME SUGGESTIONS

Because certain thread-cutting terms and the names of thread parts are used in the explanations of this assignment, it is essential that you study Fig. 51 on page 218 of your textbook. This illustration shows a Unified form of thread with all parts labeled, but these same terms apply to all forms of thread so keep this in mind as you study the setup and cutting of square and Acme threads also. As you study, pay close attention to all other illustrations in this assignment and try to remember the correct compound rest setting required to cut each form of thread as each setting varies. Note also that each form of thread requires a specially ground tool bit and how each tool bit must be ground.

When you complete the study of this assignment, answer all questions in Progress Quiz 15.

PROGRESS QUIZ 15

THE PURPOSE OF THIS QUIZ: To see how well you understand screw threads and how the lathe must be setup to perform thread-cutting operations

1. A D The threads per inch of any screw can be measured with a steel rule or a thread pitch gage. (See textbook, page 219.)

2. A D The lead of any thread is always the same as the pitch. (Page 220)

3. The included thread angle of a_______________ _______________form of thread is 60 degrees. (Page 221)

4. A D The crest of a Unified National thread may be flat or rounded. (Page 223)

5. The ground point of a Unified or V-threading tool must be checked for accuracy with a_______________ _______________. (Page 224)

(OVER)

6. A D The compound rest setting of a lathe is the same for a left- or right-hand thread. (Pages 226-227)

7. A D When cutting a thread, the lathe back gears must be engaged to reduce the speed of the lathe. (Page 227)

8. A D The setting of the gear box levers for thread-cutting operations depends upon the number of threads per inch to be cut. (Page 227)

9. A D Mineral lard oil should be used as a lubricant when threading steel shafts. (Page 228)

10. A D The included thread angle of an Acme thread is 29 degrees. (Page 233)

11. The compound rest of a lathe is set at $14\frac{1}{2}$ degrees when cutting an ____________ thread. (Page 234)

12. The width of the tool bit point for cutting a ______________ thread should be ground to $\frac{1}{2}$ the pitch of the thread. (Page 235)

(END OF QUIZ)

SUGGESTION: Check your answers by using the Answer Key in the back of the Study Guide. Review all points missed, then begin your study of Assignment 2.

MACHINE SHOP OPERATIONS AND SETUPS

ASSIGNMENT 2 and PROGRESS QUIZ 16

SUBJECT Taper and Taper Turning, Drilling, Reaming, Boring, and Internal Threading

TEXTBOOK PAGES TO STUDY 236 to 255

"YOUR AIM" in this assignment is to learn how to set up and perform the lathe operations of taper turning, drilling, reaming, boring, and internal threading.

SOME SUGGESTIONS

Because so much lathe work requires taper turning, you should make a point of trying to remember the various steps or procedures which must be followed in the setup of a lathe for taper turning. Note that taper work can be produced by three different methods. Learn what is involved in each method and what should determine your choice of the method to use for any specific job. On page 244 of your textbook, where the discussion of taper turning using the compound rest is given, a knowledge of trigonometry is required to grasp the calculations needed to find the compound rest angle setting. In most cases the shop blueprint specifies the exact angle required so these calculations are not always necessary. Because of many mathematical rules and formulas involved in the performance of all shop work, you should consider the study of some good Shop Mathematics textbook if you do not have the necessary mathematical background to understand the formulas presented in this textbook. A knowledge of Shop Mathematics is important to anyone working in a Machine Shop.

When you finish this study assignment, answer the questions in Progress Quiz 16.

PROGRESS QUIZ 16

THE PURPOSE OF THIS QUIZ: To see how well you understand how to set up and operate a lathe to perform taper turning, drilling, reaming, boring, and internal threading

1. A D A taper attachment must be used to turn a tapered shaft on a lathe. (See textbook, pages 237-238.)

2. Taper turning can be performed on a lathe when work is mounted between centers by offsetting the_______________center. (Pages 238-239)

(OVER)

3. A D The cross-feed micrometer dial can be used to determine accurately the tailstock offset needed to turn a taper. (Page 240)

4. A D The cutting edge of the tool bit when turning a taper must be set on the centerline of the work. (Page 241)

5. A D The accuracy of a machined taper can be checked only with a taper gage. (Page 241)

6. A D To machine a 60-degree included angle on a lathe center with the compound rest, the compound must be set at 30 degrees. (Page 243, Fig. 73)

7. A D Before work is center-drilled in a lathe, the ends of the work should be faced. (Page 246)

8. A D When drilling work in a lathe, the drill does not rotate as in a drill press operation. (Page 246)

9. The operation of producing an accurately-sized hole in a lathe using a tool bit is called_______________. (Page 248)

10. To determine the size of a bored hole by using a telescopic gage, the gage setting must be checked with a_____________________. (Page 251)

11. A D To cut a right-hand internal thread, the lathe compound must be set 29 degrees to the left of the cross slide centerline. (Page 252)

(END OF QUIZ)

SUGGESTION: Check your answers by using the Answer Key in the back of the Study Guide. Review points missed if necessary, then begin your study of Chapter 11, page 368 as called for in Assignment 3.

MACHINE SHOP OPERATIONS AND SETUPS

ASSIGNMENT 3 and PROGRESS QUIZ 17

SUBJECT Production Turning—Turret Lathes and Automatic Screw Machines

TEXTBOOK PAGES TO STUDY 361 to 367 and 368 to 410

"YOUR AIM" for this assignment is to gain an overall understanding of the operating principles of turret lathes and automatic screw machines and the tooling and setups used to perform specific operations.

SOME SUGGESTIONS

A careful study of the illustrations (A to F) on pages 361-367 will enable you to obtain a good understanding of turret lathe construction and the operations normally performed by such elements as the rear tool post, the square turret, and the hexagon turret. This study will prepare you for the more detailed discussions in Chapter 11 involving the function of these various lathe elements when turning, threading, drilling and boring work on a horizontal turret lathe. A careful study of this chapter will provide you with the answers to such questions as: What construction feature classifies turret lathes as a ram or saddle type? Why are these machines referred to as bar or chucking machines? Is it necessary to use a bar turner tool for all turning cuts? What operations are normally performed from the square turret, the rear tool post, or the hexagon turret? When studying this assignment, you should concentrate on learning the answers to these and similar questions in order to be proficient in the operation of turret lathes and automatic screw machines.

On the completion of this study assignment, answer the questions in Progress Quiz 17.

PROGRESS QUIZ 17

THE PURPOSE OF THIS QUIZ: To test you on your overall understanding of what must be known to operate and setup turret lathes and screw machines.

1. Stock to be machined on bar-type turret lathes is held or gripped in ________________chucks. (Page 369)

2. Turret lathes so constructed that long turning cuts are made by moving the saddle along the bedways of the machine are classified as______________-type machines. (Page 372)

(OVER)

3. A tool which is mounted in the hexagon turret for taking heavy turning cuts at high speeds is called a__________ ____________. (Page 374)

4. External and internal tapers are machined on turret lathes using a ______________ ________________. (Page 374)

5. A drill having flat sides and two cutting edges for drilling large holes is called a_____________drill. (Page 382)

6. Slide tools mounted in the hexagon turret for boring are provided with a ____________________adjustment for accurate bore size. (Page 383)

7. When reaming holes with solid reamers, mount the reamer in a ________________holder. (Page 384)

8. The use of a__________________tap when large holes are being threaded eliminates having to reverse the rotation of the lathe spindle when withdrawing the tap from the hole. (Page 387)

9. Automatic turret lathes capable of producing identical precision turned parts in large quantities are classified as______________machines. (Page 390)

10. The automatic performance of all operations on screw machines is obtained by__________. (Page 390)

11. On screw machines having vertical and cross slides, the vertical slide is normally used for_____________-___________operations. (Page 393)

12. Screw machines capable of machining more than one bar of stock at a time are classified as_______________-_______________machines. (Page 393)

13. The only screw machine having a sliding_______________which can be actuated longitudinally is the swiss-type. (Page 396)

14. The typical tolerance normally obtained in turning operations on automatic screw machines is plus and minus__________. (Page 399)

(END OF QUIZ)

SUGGESTION: Now check your answers using the Answer Key in the back of the Study Guide. Review points missed, then answer all questions in Examination 5.

CUT OFF HERE

MACHINE SHOP OPERATIONS AND SETUPS **Examination 5**

TOPIC: Engine Lathes: Cutting Tools, Setups and Operations; Production Turning: Turret Lathes and Automatic Screw Machines

Study pages 217 to 255 of Chapter 7 and pages 361 to 410 of Chapter 11.

Student's Name________________ Student Number__________

Street__________ City_________ State________ Zip Code______

1. Print the letter from the illustration next to the thread term which it represents.

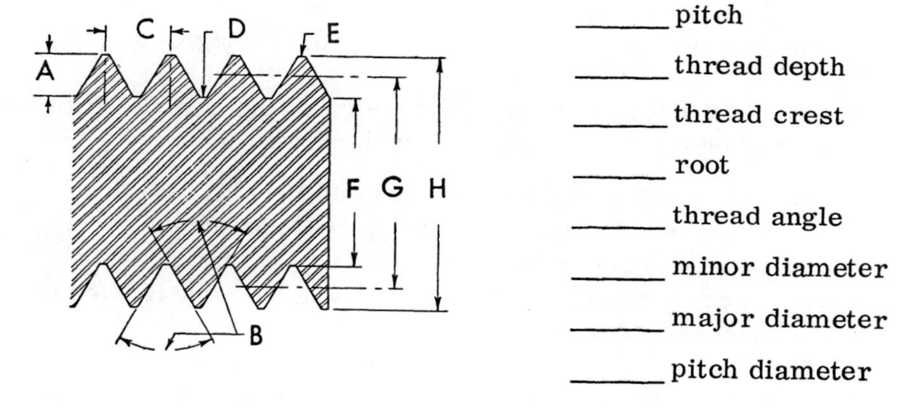

______ pitch

______ thread depth

______ thread crest

______ root

______ thread angle

______ minor diameter

______ major diameter

______ pitch diameter

2. A single thread screw having a pitch of 10 has—

   a. 8 threads per inch
   b. 10 threads per inch
   c. 9 threads per inch
   d. 6 threads per inch

3. The lead of a double-thread screw having 8 threads per inch is—

   a. .125
   b. .250
   c. .110
   d. .0625

4. The American Standard Unified thread has an included thread angle of—

   a. 30 degrees
   b. 29 degrees
   c. 60 degrees
   d. 45 degrees

(OVER)

5.

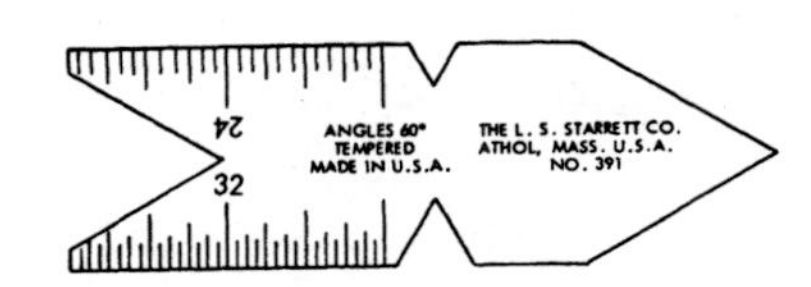

The gage shown in this illustration is called a_____________gage and is used to check the point angle of tool bits for cutting—

a. square threads
b. Unified threads
c. Acme threads
d. worm threads

6. To cut a Unified or standard V-thread, the compound rest of the lathe must be set at an angle of—

a. 15 degrees
b. 29 degrees
c. 60 degrees
d. 45 degrees

7. At what point on the face of a thread dial is the half nut lever engaged to cut an odd number of threads per inch such as 9?

a. at any of the eight graduations on the dial face
b. at any of the four numbered graduation marks
c. at only the even numbered graduation marks
d. at any graduation as long as the same point is used for each successive cut

8. The setting of the lathe gear box levers in thread-cutting operations is determined by the—

a. number of threads per inch to be cut
b. material being threaded
c. length of the thread to be cut
d. the RPM of the lathe spindle

9. A good lubricant to use in thread-cutting operations is—

a. graphite
b. white lead
c. mineral lard oil
d. water soluble oil

10. A D A lubricant is not needed when cutting threads in brass or cast iron.

11. The finish depth of an external Unified or V-thread form is most accurately measured by—

a. the three-wire system
b. a thread ring gage
c. a thread plug gage
d. a thread micrometer

CUT OFF HERE

MACHINE SHOP OPERATIONS AND SETUPS Examination 5—Continued

Student's Name____________________________ Student Number________________

12. The "best" wire size to use when measuring a 10-pitch thread (ten threads per inch) by the three-wire system is—

    a. .0437
    b. .0625
    c. .0577
    d. .031

13. The tool bit used in cutting an Acme thread must be ground to an angle of—

    a. 30 degrees
    b. 29 degrees
    c. $14\frac{1}{2}$ degrees
    d. 60 degrees

14. The thread tool in this illustration is being properly set in relation to the work using an__________thread gage. When cutting this thread, the lathe______________ must be set at an angle of $14\frac{1}{2}$ degrees.

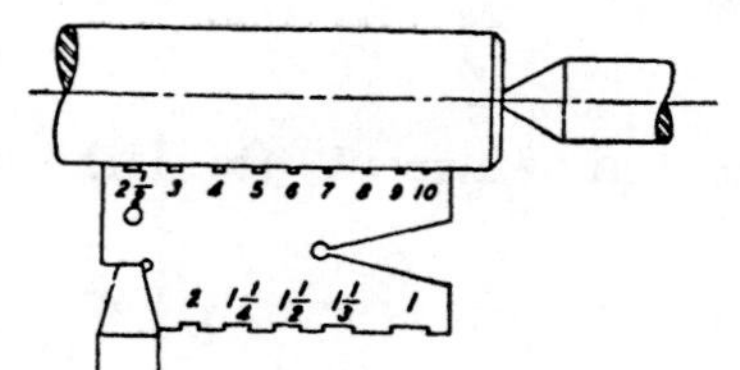

15. What would be the finish depth of a square thread having a pitch of $\frac{1}{8}$ inch?

    a. $\frac{1}{2}$ inch
    b. $\frac{1}{8}$ inch
    c. $\frac{1}{16}$ inch
    d. $\frac{3}{32}$ inch

16. A taper of approximately .625 per foot is considered to be standard for a—

    a. Jarno taper
    b. Morse taper
    c. Sellers taper
    d. Brown and Sharpe taper

17. Short or sharp angled tapers are machined using—

    a. a taper attachment
    b. the compound rest
    c. the tailstock setover method
    d. a form tool ground to the taper angle

(OVER)

18. Internal tapers are checked for accuracy using a—

a. micrometer
b. telescopic gage
c. taper plug gage
d. socket gage

19. 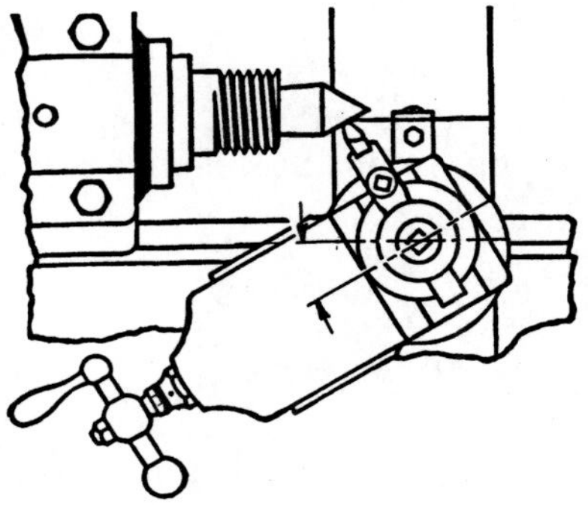 To machine the point on a lathe center as shown in the illustration, the compound rest is set at an angle of______degrees to the work centerline.

a. 30
b. 60
c. 45
d. 82

20. When an internal right-hand unified thread is being cut, the lathe compound is set—

a. parallel to the work axis
b. 29 degrees to the left of the cross-slide centerline
c. 90 degrees to the work axis
d. 29 degrees to the right of the cross-slide centerline

21. To thread small diameter holes on a lathe, they must be____________.

NOTE: The following questions apply to turret lathes and screw machines.

22. On bar-type turret lathes, work to be machined is gripped or held by—

a. collet chucks
b. universal chucks
c. faceplates
d. independent chucks

23. Turret lathes on which long turning cuts are made by movement of the saddle along the bedways of the machine are classified as—

a. saddle-type turret lathes
b. ram-type machines
c. carriage-type machines
d. automatic screw machines

Student's Name________________________ Student Number________________

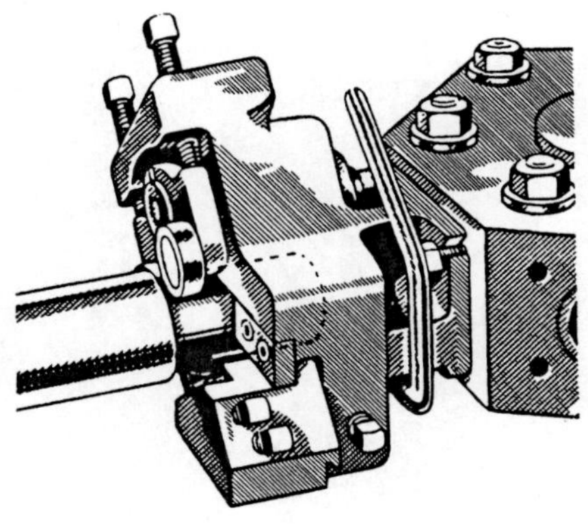

24. The tool shown in this illustration is called a________ ____________. To obtain a highly burnished surface of the turned shaft, the rolls should be set _____________the cutting tip of the single-point tool.

25. Taper turning on a turret lathe requires the use of a taper attachment. The turning cut is then taken from the—

   a. square turret
   b. hexagon turret
   c. vertical tool slide
   d. rear tool post

26. The drill shown in this illustration is called a—

   a. core drill
   b. spade drill
   c. pilot drill
   d. counterboring drill

27. When work from the hexagon turret is being bored, the boring bar holding the tool bit must be mounted in a____________ ____________.

28. On single-spindle screw machines, stock cutoff operations are normally performed from the_______________tool slide.

29. The movement of the various slides as well as the feeding of the stock is entirely automatic on screw machines and is obtained by________action.

30. A distinguishing feature of Swiss-type screw machines is that they are equipped with a____________spindle which can be actuated longitudinally.

(END OF EXAMINATION)

CUT OFF HERE

SECTION 6 — Study Guide for MACHINE SHOP OPERATIONS AND SETUPS

TEXTBOOK CHAPTER COVERED

CHAPTER 8 — Shaper Construction, Setups, and Operations

CONTENTS OF SECTION 6

Assignment 1 and Progress Quiz 18 — Study textbook pages 256 to 278, shaper construction and the setups used to perform various operations on this machine.

MEMORY JOGGER 2

EXAMINATION 6 — Based on textbook pages 256 to 278

WHAT THE STUDY OF SECTION 6 WILL DO FOR YOU

To help you expand your knowledge of the basic machine tools in the machine shop, this section of the Study Guide will test you on your understanding of how to operate and set up a shaper, to perform machining operations such as horizontal, vertical, and angular surface cutting, keyway cutting, slotting, serrating and dovetail cutting. The primary function of the shaper is to produce a flat machined surface, but note that the shaper can also be used for slotting and angle cutting if and when required. You will learn how to operate and set up a shaper to perform the work expected from it.

Now begin your study of Assignment 1.

MACHINE SHOP OPERATIONS AND SETUPS

ASSIGNMENT 1 and PROGRESS QUIZ 18

SUBJECT Shaper Construction, Setups, and Operations

TEXTBOOK PAGES TO STUDY 256 to 278

"YOUR AIM" for this assignment is to gain the knowledge and understanding necessary to help you become skillful in the operation and setting up of a shaper.

SOME SUGGESTIONS

To understand fully how to set up a shaper to perform any specific operation you may be asked to do, you must first have a working knowledge of the machine tool itself. As you study, try to remember, not only the name but also the function of all important parts of this machine as well as how these parts are adjusted. It is also suggested that you refer to all illustrations in the assignment so that you will learn to identify these parts by sight and will know where they are located.

When you finish this study assignment, answer all questions in Progress Quiz 18.

PROGRESS QUIZ 18

THE PURPOSE OF THIS QUIZ: To see how well you understand the operating principles of a shaper.

1. A D All work machined on a shaper must be securely mounted or held in a shaper vise. (See textbook, page 258.)

2. So that the depth of any cut can be accurately determined, the tool head feed screw is equipped with a____________________collar graduated in thousandths of an inch. (Page 259)

3. A D Shaper size is determined by the maximum length of the ram stroke in inches. (Page 260)

4. On a____________________shaper, oil flow from a high-pressure pump is used to power the ram. (Page 262)

5. A D Vertical shapers are equipped with a table having a cross feed, a longitudinal feed, and a circular feed. (Page 264)

(OVER)

6. To properly seat work in a shaper vise for machining, the work should be supported on_______________. (Page 267)

7. Vee blocks are commonly used as work holding devices for___________ shafts which require shaping. (Page 267)

8. The_____________of the ram stroke is regulated by turning the stroke-adjusting shaft. (Page 268)

9. A D Internal slots or keyways cannot be machined on a horizontal shaper. (Page 273, Fig. 24)

10. To obtain an equal spacing of splines around the circumference of a shaft by shaping, the shaft is mounted between_____________centers. (Page 274)

11. A D An angle cut can be taken by horizontal shaping if the part is properly mounted in a vise. (Page 276)

(END OF QUIZ)

SUGGESTION: Check your answers by using the Answer Key in the back of the Study Guide. Review all points missed if necessary, then answer all questions in Memory Jogger 2 to see how well you can recall points studied in the last three sections of the Study Guide.

MACHINE SHOP OPERATIONS AND SETUPS

MEMORY JOGGER 2

THE PURPOSE OF THIS QUIZ: To determine how well you can recall the important points you have studied in Chapters 6, 7, 8, and 11 of your textbook.

Note: The following questions relate to engine lathes.

1. Square or irregular shaped stock for turning is usually mounted in an ____________________ chuck. (Page 165)
2. A device which is fastened to the headstock end of work to be turned between centers is called a_____________ __________. (Page 171)
3. Previously drilled and reamed parts can be turned between centers by mounting such parts on a__________________. (Page 172)
4. The angles ground into the face or top surfaces of lathe tool bits are referred to as_____________angles. (Page 183)
5. The angles which are ground on the sides of a tool bit are called ____________________angles. (Page 184)
6. When grinding carbide tool bits for finish cuts, use a________________ wheel. (Page 199)
7. To obtain accurate centering of stock mounted in an independent chuck, a______________ ______________is used. (Page 200)
8. Which of the following lathe operations requires that the cutting edge of a tool bit be placed exactly on the work centerline? (Page 202)

   a. boring b. facing c. turning
9. To check the point angle of tool bits for cutting Unified threads a ________________gage is used. (Page 225)
10. When cutting an external Unified thread, set the lathe________________ 29 degrees to the right of the cross slide centerline. (Page 226)

Note: The following questions relate to shapers and shaper work.

11. Shapers equipped with a table that can be swiveled for angle cutting are referred to as____________________shapers. (Page 259)
12. Rotary tables for circular cuts are standard equipment on a___________ shaper. (Page 264)
13. To properly seat work in a shaper vise for machining, the work is supported on____________________. (Page 267)

(OVER)

14. The work holding device commonly used when machining a keyway in a cylindrical shaft on a shaper is— (Page 267)

    a. an angle plate b. a vise c. a Vee block

15. To obtain an equal spacing of splines around the circumference of a shaft by shaping, the shaft is mounted between______________centers. (Page 274)

Note: The following questions are based on your study of <u>turret</u> <u>lathes</u>.

16. Turret lathes equipped with spindles which can be fitted with a universal 3-jaw chuck are referred to as______________machines. (Page 370)

17. Long turning cuts are made on___________-type turret lathes by longitudinal movement of the saddle along the bedways of the machine. (Page 372)

18. A tool which is mounted in the hexagon turret when taking heavy turning cuts at high speeds is called a________ ____________. (Page 374)

19. When machining internal or external tapers on a turret lathe, you must use a____________ _________________. (Page 374)

20. A flat sided drill which has two cutting edges and which is used for drilling large holes is called a_____________drill. (Page 382)

21. To eliminate having to reverse the machine spindle to withdraw a tap from a threaded hole, a_________________tap should be used. (Page 387)

22. Automatic turret lathes capable of producing large quantities of identical precision turned parts are referred to as____________machines. (Page 390)

23. All moving elements used in a machining operation on a screw machine are positioned for cuts automatically by_________action. (Page 390)

24. On screw machines having cross and vertical slides, cutting off operations are usually performed from the______________slide. (Page 393)

25. Screw machines capable of machining more than one bar of stock at a time are classified as______________-______________machines. (Page 393)

26. Tolerances normally obtained in screw machine turning operations are plus and minus_________. (Page 399)

(END OF QUIZ)

Now check your answers using the Answer Key in the back of the Study Guide. Review all points missed; then answer all questions in Examination 6.

Examination 6

MACHINE SHOP OPERATIONS AND SETUPS

TOPIC: Shaper Construction, Setups, and Operations

Based on pages 256 to 278 in the textbook

Student's Name ______________________ Student Number ________________

Street ______________ City ___________ State ___________ Zip Code __________

1. The shaper part which guides the horizontal table movement at right angles to the ram stroke when the power feed is engaged is called a ___________.

2. Precise settings for depth of cut are obtained by an adjustment of the _________________ collar on the shaper tool head.

3. Shapers equipped with universal rather than the conventional type table are preferred when performing operations such as—

   a. vertical shaping
   b. angle cutting
   c. slotting
   d. form cutting gear teeth

4. The cutting action of a shaper occurs only on the ______________ stroke of the ram.

5. The hinged part on the front face of a shaper tool head on which the tool post is mounted is called a ______________ __________.

6. The clapper box is usually set vertically when taking—

   a. roughing cuts
   b. angular cuts
   c. horizontal cuts
   d. vertical cuts

7. The factor to be considered in determining shaper size is the—

   a. overall height of the machine
   b. the table size with which the machine is equipped
   c. the maximum table feed movement
   d. the maximum stroke length of the ram

8. The large gear used in crank-type shapers to change rotary motion to reciprocating motion is commonly referred to as a _________ _________.

9. A D Hydraulic shapers are provided with a greater range of speeds and feeds than crank-type shapers.

(OVER)

CUT OFF HERE

10. Vertical shapers differ from horizontal shapers in that they have a ________ ram and are equipped with a ________ table.

11. When grinding single point tools for shaper use, what degree of side clearance is normally provided?

    a. 5 degrees
    b. 2 degrees
    c. 15 degrees
    d. 8 degrees

12. To prolong the life of shaper tools after they are ground, they should be—

    a. lapped
    b. sanded
    c. stoned
    d. hardened

13. The smoothly finished square or rectangular bars used as a workseat support when shaping work in a vise are called ________.

14. What work holding device would be recommended for use when shaping a keyway in a cylindrical shaft?

    a. a Vee block
    b. an angle plate
    c. a dividing head
    d. a shaper vise

15. When setting up a shaper, the ram stroke must be adjusted for both ________ and ________ over the work.

16. The adjustment of the ram-positioning shaft on a shaper is done with the ram—

    a. stopped at the end of the back or return stroke
    b. in motion on the return stroke
    c. stopped midway through the cutting stroke

17. To machine a piece of work $5\frac{1}{2}$ inches long on a shaper, the minimum stroke length which can be used is—

    a. $5\frac{3}{4}$ inches
    b. $5\frac{1}{2}$ inches
    c. $6\frac{1}{4}$ inches

MACHINE SHOP OPERATIONS AND SETUPS Examination 6—Continued

Student's Name ______________________________ Student Number ____________________

CUT OFF HERE

18. This illustrates the use of a ______________ ______________ to check the shaper head when a ______________ cut is taken, using the tool head feed.

19.

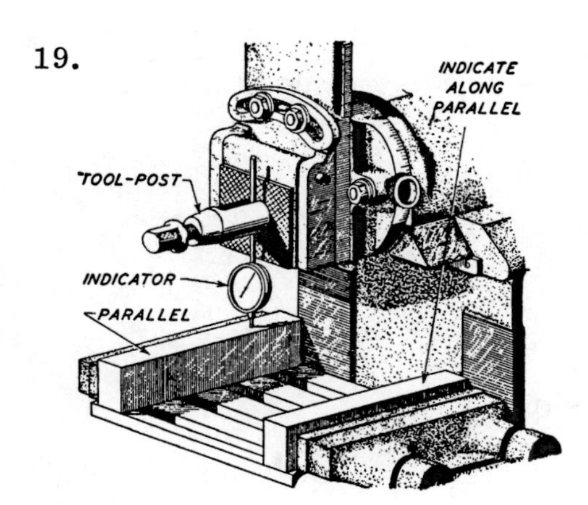

This illustrates the correct method of checking the workseat of a shaper vise for being parallel to the ______________ of the __________.

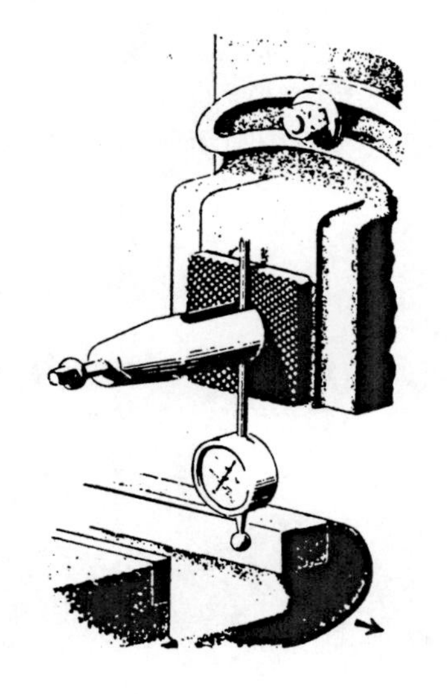

20. This illustrates the use of a __________ ______________ to check the vise jaws for being at right angles to the ______________ of the ram.

(OVER)

21. To shape splines in a shaft which must be accurately spaced, the work is mounted—

    a. in a shaper vise
    b. between indexing centers
    c. in V-blocks
    d. in a special fixture

22. A D When taking a vertical cut using the hand feed on the tool head, the clapper box must be set at an angle from the vertical position of the head.

23. A D It is not always necessary to set the tool head at an angle when taking an angular cut.

24. As a safety precaution when operating a shaper not equipped with a_______ _____________, the operator should wear safety goggles.

25. Safe operating practice requires that the shaper be_____________before any changes of machine adjustment or in the work setup are made.

(END OF EXAMINATION)

SECTION 7 — Study Guide for MACHINE SHOP OPERATIONS AND SETUPS

TEXTBOOK CHAPTERS COVERED

Chapters 9 and 10 — Milling Machines: Types, Construction, Accessories, Cutting Tools, Setups and Operations

CONTENTS OF SECTION 7

Assignment 1 and Progress Quiz 19 — Study textbook pages 279 to 306, milling machine types, construction accessories and attachments.

Assignment 2 and Progress Quiz 20 — Study textbook pages 307 to 327, milling machine cutters and the basic machine tool adjustments to observe when making setups.

Assignment 3 and Progress Quiz 21 — Study textbook pages 327 to 360, the cutting tools and accessories used, and the setups needed to perform all milling machine operations.

EXAMINATION 7 — Based on pages 279 to 360 of your textbook

WHAT THE STUDY OF SECTION 7 WILL DO FOR YOU

A complete working knowledge of a milling machine and how to set up and perform operations on it in a skillful manner is most essential in the training of any machinist or tool and die maker. Before you can develop this skill, you must understand the construction of the different types of milling machines, the accessories and attachments which are used to perform specific operations, the many different types of milling cutters used and how these cutters are mounted, the proper mounting of work and how to adjust the machine to obtain the correct speed and feed in the performance of different operations. These are the many things you will learn in the study of this section.

Now begin your study of Assignment 1.

MACHINE SHOP OPERATIONS AND SETUPS

ASSIGNMENT 1 and PROGRESS QUIZ 19

SUBJECT Milling Machines: Types, Construction, Accessories, and Attachments

TEXTBOOK PAGES TO STUDY 278 to 307

"YOUR AIM" Before you can learn to operate a milling machine, you must understand the names and function of the important parts of the machine as well as the accessories and attachments which can be used in performing certain operations.

SOME SUGGESTIONS

Before you begin to read and study this assignment, you should refer to and study the illustrations, Figures 7 to 11. Doing so will help you to see quickly what a milling machine looks like and what the names of the important parts are as well as the different controls used to move these parts by hand or power. After you study the illustrations, you will better understand the written material, explaining the function and names of milling machine parts, later in the assignment.

When you finish this study assignment, answer the questions in Progress Quiz 19.

PROGRESS QUIZ 19

THE PURPOSE OF THIS QUIZ: To see how well you understand milling machine types and construction and the accessories and attachments used

1. A D When performing peripheral milling operations, the milling cutters are mounted on an arbor. (See textbook, page 279.)

2. To prevent springing the arbor of a horizontal milling machine, the arbor nut should not be tightened until the ______________ ______________ is in place. (Page 282)

3. A D Face milling operations can be performed only on a vertical milling machine. (Page 284)

4. The position of the ______________ is what classifies a milling machine as being horizontal or vertical. (Page 287)

5. The depth of cut in horizontal milling operations is obtained by raising the ____________ which supports the table saddle. (Page 291)

(OVER)

6. The feed screws of a milling machine table are provided with__________ dials to permit accurate positioning of the table. (Page 291)

7. A D Proper positioning of milling cutters on an arbor is obtained with spacing collars. (Page 295)

8. A D Vertical milling attachments are for use on horizontal milling machines. (Page 299)

9. The attachment used on milling machines for indexing is called a________ __________. (Page 300)

10. A D Forty turns of the index crank are required when indexing, to rotate the workpiece one complete turn. (Page 301)

11. A D The Brown and Sharpe dividing head is equipped with a set of three index plates which cover a hole circle range of 15 to 49. (Page 301)

12. A D When a partial turn of the work is needed, the sector arms should be set for the number of holes needed for indexing. (Page 302)

13. A D It is important in indexing to see that the workpiece is clear of the cutter before the work is indexed for each cut. (Page 302)

14. A D To disengage the index crank for indexing, first withdraw the index pin from the index plate hole. (Page 302)

15. A D When indexing work by the differential method of indexing, a universal dividing head must be used. (Page 303)

16. The first step in calculating the correct indexing needed for a specific job is to divide the constant______by the number of divisions required. (Page 304)

17. A D To index a job requiring five evenly-spaced divisions using a Brown and Sharpe dividing head, the number 2 or 3 plate could be used. (Pages 301-304)

18. A D When performing angular indexing, it is important to remember that one complete crank turn is required to index the work 10 degrees. (Page 304)

(END OF QUIZ)

SUGGESTION: Now check your answers by using the Answer Key in the back of the Study Guide. Review all points missed, then begin your study of Assignment 2.

MACHINE SHOP OPERATIONS AND SETUPS

ASSIGNMENT 2 and PROGRESS QUIZ 20

SUBJECT   Milling Machine Cutters and Machine Adjustments Needed in Setups

TEXTBOOK PAGES TO STUDY   307 to 327

"YOUR AIM" for this assignment is to learn the many types of milling cutters used in milling operations and the machine adjustments to make when setting up work for milling.

SOME SUGGESTIONS

It is important in this assignment that you learn the names and types of milling cutters used in milling operations. This can be understood when you realize that some cutters can be used to perform several operations, but that other operations can be performed by using only one special kind of cutter. The selection of the proper cutter for the job is very important in milling operations. It is also important to remember the eleven steps, listed on pages 307 and 308 of your textbook, which must be observed when setting up the milling machine for any operation.

When you complete the study of this assignment, answer all questions in Progress Quiz 20.

PROGRESS QUIZ 20

THE PURPOSE OF THIS QUIZ: To see how well you can identify and name the different types of milling cutters used and what machine adjustments are needed in setting up a mill.

1. A D Tungsten Carbide tipped milling cutters retain their hardness and sharpness at higher temperatures better than cutters made of high-speed steel. (See textbook, page 308.)

2. A D Plain milling cutters can have straight or helical teeth and are used in milling flat surfaces. (Page 310)

3. A D The size of the helix angle of a helical cutter determines whether it should be used in light or heavy cutting operations. (Page 310)

4. A D A cutter having both peripheral and side cutting teeth is called a plain milling cutter. (Pages 310-311)

(OVER)

5. The cutters designed for straddle milling operations are called_______ milling cutters. (Page 311)

6. The cutting action of a__________milling cutter is the same as an end mill cutter. (Page 312)

7. A D A dovetail milling cutter is really just an end mill cutter having a special form. (Page 315)

8. In horizontal milling operations, the__________________feed is used to move the workpiece into or away from the cutter. (Page 317)

9. A D The direction of work feed and cutter rotation are the same when climb milling. (Page 320)

10. A D Climb milling should not be used in machining cast iron because the surface scale dulls the cutter teeth. (Page 320)

11. A D To obtain fine finish cuts in milling, the speed of the machine should be increased and the feed decreased. (Page 326)

(END OF QUIZ)

SUGGESTION: Now check your answers by using the Answer Key in the back of the Study Guide. Review all points missed, then begin your study of Assignment 3.

MACHINE SHOP OPERATIONS AND SETUPS

ASSIGNMENT 3 and PROGRESS QUIZ 21

SUBJECT How To Setup and Perform Specific Milling Operations

TEXTBOOK PAGES TO STUDY 327 to 360

"YOUR AIM" for this assignment is to learn the correct steps to follow in setting up a milling machine to perform operations such as slab milling, face milling, end milling, side milling, straddle milling, and gear milling.

SOME SUGGESTIONS

As you study this assignment, try to remember the important steps to follow in setting up the milling machine for each operation discussed. This involves the points to observe in mounting the arbors, the cutters, the correct selection of cutter type, the proper selection of work-holding devices, the mounting of the work, the table adjustments to be made, and the correct selection of speeds and feeds. Your skill in the operation of a milling machine is entirely dependent upon how well you remember and can apply what you will read about and learn in this assignment.

When you finish this study assignment, answer all questions in Progress Quiz 21.

PROGRESS QUIZ 21

THE PURPOSE OF THIS QUIZ: To see how well you have learned the important steps to be followed when setting up a milling machine for specific operations

1. A D Cutters used for milling flat horizontal surfaces must be mounted on an arbor. (See textbook, page 327.)

2. A style "C" arbor must be used for mounting____________-__________ mill cutters. (Page 332)

3. The________________of the cutter is an important factor to consider in the horizontal milling of deep slots or channels, to prevent interference between the workpiece and arbor. (Page 334)

(OVER)

4. A quick and accurate way to mill a perfect square on the end of a round shaft is to mount the shaft in a ______________ ___________ so the shaft can be indexed. (Page 334)

5. A D In straddle-milling operations, the spacing of the cutters on the arbor determines the width of the part milled. (Page 335)

6. A D Bevel gears cannot be used to transmit power or motion between parallel shafts. (Page 340)

7. A D The pitch diameter of a spur gear, having 35 teeth and a diametral pitch of 7, is 5 inches. (Page 344)

8. The accurate spacing of teeth in a gear blank requires the use of a ________________ ___________. (Page 346)

9. The cutting of helical gears requires the use of a_______________ milling machine. (Page 349)

10. A D In helical milling, the gear blank is rotated by engaging the table feed. (Page 349)

11. A gear-tooth vernier caliper is used to measure the________________ and chordal thickness of gear teeth. (Pages 355-356)

(END OF QUIZ)

SUGGESTION: Now check your answers by using the Answer Key in the back of the Study Guide. Review all points missed, then answer all questions in Examination 7.

MACHINE SHOP OPERATIONS AND SETUPS **Examination 7**

TOPICS: Milling Machine Construction, Cutting Tools, and Accessories

Study pages 279 to 360 in the textbook

Student's Name________________ Student Number____________

Street__________ City__________ State__________ Zip Code______

1. A milling machine which has a table that can be swiveled and set at any angle to the column face is called a—

   a. plain knee-and-column type
   b. universal knee-and-column type
   c. bed-type

2. The only milling arbor which does not require the use of an arbor support is the—

   a. style A arbor
   b. shell-end mill arbor
   c. style B arbor

3. To index work mounted on a dividing head one complete turn, the index crank must be turned—

   a. forty complete turns
   b. twenty complete turns
   c. ten complete turns

4. The number of index crank turns needed after each cut to mill eight equally spaced slots in a round shaft is—

   a. four
   b. seven
   c. five

5. A D The plain dividing head can be used for direct and differential indexing.

6. Only a universal dividing head can be used to perform a milling operation by—

   a. plain indexing
   b. direct indexing
   c. differential indexing

(OVER)

CUT OFF HERE

7. One complete turn of the index crank when indexing work by degrees will turn the workpiece—

   a. 9 degrees
   b. 15 degrees
   c. 7 degrees

Directions: Identify the various milling cutters by writing the name of the cutter on the lines provided under each illustration.

8. ____________________

9. ____________________

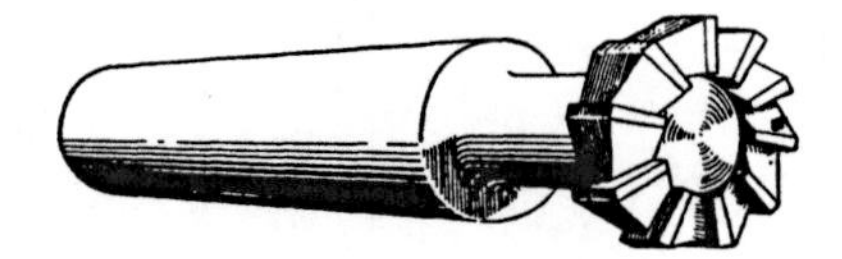

10. ____________________

11. ____________________

Student's Name ______________________________ Student Number ______________

12. In the illustration shown, the climb milling method of cutter rotation relative to work feed is shown in Fig.____

Fig.___shows the conventional method.

Fig. A

Fig. B

13. Because the hard scale will dull the cutter teeth, the climb cutting method of cutter rotation should never be used when milling—

    a. tool steel
    b. magnesium
    c. cast iron

14. The rigidity of the milling machine and the horsepower rating must be adequate when a milling machine operation must be performed using—

    a. carbide-tipped milling cutters
    b. cast-alloy milling cutters
    c. high-speed steel milling cutters

15. When all factors that determine the proper choice of cutting speeds to use in milling are considered, the cutting tool material capable of withstanding the highest cutting speeds is the—

    a. stellite tipped cutters
    b. tungsten-carbide tipped cutters
    c. super high-speed cutters

16. A D A rule to keep in mind when mounting cutters on an arbor is that they should always be mounted as close to the column of the machine as possible. (Page 328)

(OVER)

CUT OFF HERE

17. This illustrates how a__________ ______________is mounted and used to align the solid jaw of a milling vise ______________to the face of the milling machine column.

18. 

This illustrates a setup commonly employed when it is required to mill two surfaces at_____degrees to each other in one pass of the cutter. The cutter used in this case is a______________ ________ milling cutter.

19. 

The milling operation illustrated here is called______________mill-ing. The proper__________of the cutters on the arbor is very im-portant in this operation as this will determine the width of the part milled.

20. The measurement from center to center of adjacent gear teeth taken on the arc of the pitch circle, is called the—

a. circular pitch
b. chordal pitch
c. diametral pitch

Student's Name______________________________ Student Number____________

21. Write the correct letter from the illustration below, to identify each gear-tooth term listed.

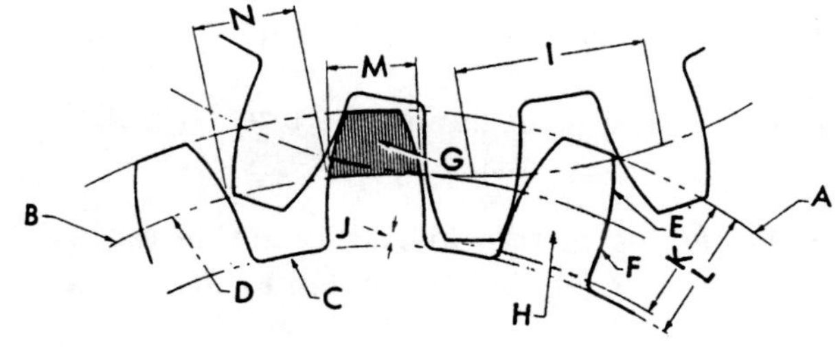

Chordal Tooth Thickness ______

Pitch Circle ______

Tooth Face ______

Tooth Flank ______

Addendum ______

Dedendum ______

22. The tooth addendum of a gear having a diametral pitch of 4 is—

    a. .250
    b. .125
    c. .156

23. What would be the proper number of form cutter to use in milling a gear with 36 teeth?

    a. No. 6 cutter
    b. No. 3 cutter
    c. No. 5 cutter

24. It is important when milling gear teeth that the mandrel used to hold the gear blank is mounted, so that its large end is toward the____________ of the dividing head.

25. Before the first cut in milling teeth in a gear blank is taken, the cutter must be____________over the gear blank.

26. Helical milling operations can be performed only on a—

    a. plain horizontal milling machine
    b. vertical milling machine
    c. universal milling machine

CUT OFF HERE

(OVER)

27. The circumference of the gear blank must be divided by the lead of the helix in helical milling to determine the —

    a. angle setting of the machine table
    b. proper speed to use
    c. proper feed and depth of cut needed

28. A D A standard or plain dividing head cannot be used in milling helical gears.

29. A D Except for the angle setting of the milling machine table, the setup for milling a right- or left-hand helix is the same.

30. It is important in milling a helical gear to center the__________over the mandrel before the table is set to the helix angle.

31. 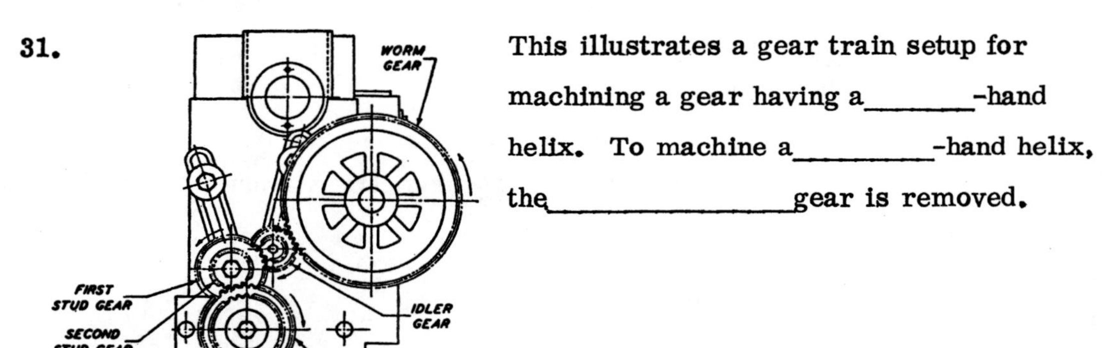

This illustrates a gear train setup for machining a gear having a_______-hand helix. To machine a_________-hand helix, the________________gear is removed.

32. A D The teeth of gear racks can be indexed for accurate spacing only by using a special rack-cutting attachment.

33. A D When measuring gear teeth for size with a gear-tooth vernier, the reading for chordal tooth thickness is obtained from the horizontal scale of the caliper.

(END OF EXAMINATION)

SECTION 8 — Study Guide for MACHINE SHOP OPERATIONS AND SETUPS

TEXTBOOK CHAPTER TO STUDY

CHAPTER 12 — Grinding Machines: Surface, Cylindrical, and Internal

CONTENTS OF SECTION 8

Assignment 1 and Progress Quiz 22 — Study textbook pages 411 to 449, the operating principles of all types of Grinders and their Setups.

EXAMINATION 8 — Based on pages 411 to 449 of the textbook

WHAT THE STUDY OF SECTION 8 WILL DO FOR YOU

In the study of this section, you will learn about the machine tool called a grinder. The primary function of this machine is to produce a finer finish or closer size on workpieces which are machined first on a lathe, milling machine, or a shaper. Occasionally parts are completely ground from rough stock without previous machining, but this is possible only in a few cases. Because grinding is the last or final operation on piece parts, you can understand why it is important that you learn or understand how to setup and operate this machine tool in a most efficient manner. Ground parts require extreme accuracy in size and surface finish, and if these results are not obtained in this final operation the entire part may have to be scrapped. This is costly as it means all time required in previous machining operations has been wasted. The setup and operation of various types of grinders is what you will learn in this section.

Now begin your study of Assignment 1.

MACHINE SHOP OPERATIONS AND SETUPS

ASSIGNMENT 1 and PROGRESS QUIZ 22

SUBJECT Grinding Machines: Surface, Cylindrical, Internal, and Tool Grinders

TEXTBOOK PAGES TO STUDY 411 to 449

"YOUR AIM" for this assignment is to acquire the knowledge needed to operate and setup grinders such as the surface types, the center and centerless types of cylindrical grinders, and the internal grinders.

SOME SUGGESTIONS

An important point to keep in mind as you study this section to learn how to setup and operate the various grinders discussed in this chapter, is that most grinding work is performed on parts previously machined on machine tools studied in earlier chapters. Blueprints only specify the finish size of all parts, so when you turn a shaft on a lathe which calls for finish grinding, you must remember not to turn the shaft to the dimension given on the print. The shaft must be left oversize to allow for the stock to be removed by finish grinding. The amount of stock to leave and what determines this is what you should try to remember. Also note the construction differences and operational controls provided on the various types of grinders as well as the methods used in setting up these grinders to perform specific grinding operations. Other important points to keep in mind are what determines the correct selection of grinding wheel used and how the wheel must be dressed and mounted.

When you finish this study assignment, answer the questions in Progress Quiz 22.

PROGRESS QUIZ 22

THE PURPOSE OF THIS QUIZ: To see how well you understand the operation and setup of each grinder discussed in this assignment

1. The grinder which is used to produce flat ground surfaces is called a ____________________ grinder. (See textbook, page 411.)

2. The depth of cut on a surface grinder is regulated by raising or lowering the______________________. (Page 413)

(OVER)

3. Most surface grinders are provided with a________________chuck for holding the work to be ground. (Page 414)

4. A D All work to be ground on a cylindrical grinder must be held or mounted on centers similar to those used in lathe work. (Page 418)

5. The operation of sharpening a grinding wheel is called wheel____________. (Page 420)

6. To form-grind work accurately on a cylindrical grinder, the____________ __________method of grinding must be used. (Page 428)

7. A D In centerless grinding operations, the work is made to rotate by the regulating wheel. (Page 428)

8. A D The regulating wheel and the grinding wheel must always rotate at the same speed in centerless grinding operations. (Page 428)

9. A D Internal grinding cannot be performed on a centerless grinder as there is no provision for chucking the work. (Page 431)

10. A D When sharpening milling cutters by grinding, the width of the land must be reduced, if it exceeds 1/16 inch, by grinding a secondary clearance angle. (Pages 441-442)

(END OF QUIZ)

SUGGESTION: Now check your answers to these quiz questions by using the Answer Key in the back of the Study Guide. Review all points missed by referring to textbook pages given after each question, then answer all questions in Examination 8.

CUT OFF HERE

MACHINE SHOP OPERATIONS AND SETUPS

Examination 8

TOPICS: Grinding Machines: Types—Construction—Operation and Setup

Study pages 411 to 449 in the textbook

Student's Name____________________ Student Number______________

Street____________ City____________ State____________ Zip Code______

1. Grinding machines equipped with rotating worktables are classified as—

   a. tool or cutter grinders
   b. cylindrical grinders
   c. surface grinders

2. Precision work cannot be produced on a surface grinder having worn ______________bearings.

3. The grinding wheels employed in surface grinding operations are classified as—

   a. plain grinding wheels
   b. cup wheels
   c. dished or saucer wheels

4. When a coolant is used in grinding operations, the coolant flow must be directed to the contact area between the__________and the__________.

5. A D In all cylindrical grinding operations, the workpiece must be mounted or held between centers so that the work can rotate.

6. In center-type cylindrical grinding operations when the work is mounted between centers, the work is made to rotate by—

   a. using a live or rotating center in the headstock spindle
   b. the frictional drive of a regulating wheel
   c. the same general method used to rotate work mounted between centers on a lathe

7. A D The workpiece and the grinding wheel must rotate in the same direction when grinding work cylindrically.

F4-4

(OVER)

8. When grinding wheels become loaded or glazed, they may be dressed, to restore sharpness, with a —

   a. pumice stone
   b. tool steel dresser
   c. table type diamond-wheel dresser

9. When grinding long slender work on a cylindrical grinder, use__________ ______________to prevent the work from springing away from the grinding wheel.

10. Depending upon their construction, center-type cylindrical grinders are classified as being_________________or______________________grinders.

11. A center-type cylindrical grinder that permits swiveling the wheelhead and headstock at an angle to the table ways is called a_________________ grinder.

12. A D In cylindrical grinding operations, the work is always rotated at a much slower speed than that of the grinding wheel.

13. A D The peripheral speed in feet per minute of a 6-inch diameter grinding wheel rotating at 2, 865 r.p.m. is 4, 500.

14. The type of grinding which involves feeding the grinding wheel into the workpiece is called _________________.

15. A D Centerless grinding is classified as a cylindrical grinding operation.

16. A slight taper to be ground the full length of a long shaft mounted between centers on a universal cylindrical grinder is accomplished by—

    a. offsetting the footstock
    b. swiveling the table on its base
    c. swiveling the wheelhead

17. To grind the 60-degree angle on the end of a machine center point on a universal cylindrical grinder, the—

    a. table is swiveled to 30 degrees
    b. headstock is swiveled to 60 degrees
    c. headstock is swiveled to 30 degrees

Student's Name______________________________ Student Number_______________

18. Plunge-cut grinding is accomplished on a cylindrical grinder by—

    a. traversing the table
    b. using the table cross feed
    c. feeding the grinding wheel into the work while the table remains stationary

19. The correct height setting of the work in________________grinding operations is obtained by the setting of the work rest blade.

20. In centerless grinding operations, the work is rotated and fed past the face of the grinding wheel by the______________________ ________________.

21. A D In centerless grinding operations, the workpiece rotates in an opposite direction to the rotation of the grinding and regulating wheels.

22. Holes in parts which have been hardened by heat treating can be finished to accurate size only by—

    a. drilling
    b. boring
    c. internal grinding

23. The surface speed of the smaller grinding wheels used for internal grinding operations ranges from—

    a. 50 to 100 surface ft. per minute
    b. 1000 to 3000 surface ft. per minute
    c. 4000 to 6000 surface ft. per minute

24. A D On a work-rotating internal grinder the wheel and the work centerlines are in the same vertical plane.

25. When grinding the teeth of a milling cutter on a tool grinder, it is important to provide_________________for the cutting edges.

26. A D An aluminum-oxide grinding wheel is a better wheel to use, when grinding any of the hard steels, than the silicon carbide wheel.

(END OF EXAMINATION)

CUT OFF HERE

SECTION 9 — Study Guide for MACHINE SHOP OPERATIONS AND SETUPS

TEXTBOOK CHAPTERS TO STUDY

CHAPTERS 13 and 14 — Steel and Its Alloys; Heat Treating

CONTENTS OF SECTION 9

Assignment 1 and Progress Quiz 23 — Study pages 450 to 467, the various steels and steel alloys, their characteristics and uses.

Assignment 2 and Progress Quiz 24 — Study pages 468 to 487, heat-treating processes, equipment and hardness testing.

MEMORY JOGGER 3

EXAMINATION 9 — Based on pages 450 to 487 of your textbook

WHAT THE STUDY OF SECTION 9 WILL DO FOR YOU

To be recognized as a fully competent Machinist or Tool and Die Maker, requires not only that you know how to set up and operate the various machine tools you have studied in the preceding chapters of the textbook, but also that you understand the characteristics of the various metals to be machined. This is very important because unless you understand the various classifications of steel and what qualities each steel may possess, you may select the wrong steel for a specific use. As a result, the cutting tool, the drill jig, or die you make will not withstand the work it was designed to perform. Knowing how to select the proper steel to suit a specific requirement is an important part of the job and is what you will learn in the study of this section.

MACHINE SHOP OPERATIONS AND SETUPS

ASSIGNMENT 1 and PROGRESS QUIZ 23

SUBJECT Steel and Its Alloys, Types and Characteristics

TEXTBOOK PAGES TO STUDY 450 to 467

"YOUR AIM" for this assignment is to acquire the knowledge which will enable you to select the proper type of steel to suit a specific requirement or use.

SOME SUGGESTIONS

An important part of the study for this assignment is to learn the two coding systems used to classify steels. Both the SAE and the AISI coding methods are commonly used by industry, and you must understand the exact meaning of each letter or number in each system to be able to identify the composition of a steel, whether it is a high or low carbon or an alloy steel, by its code number. Table II on pages 461 and 462 of the textbook should be studied carefully as the special characteristics of many types of steels are not only described, but also the common uses of the steels are named. This should prove to be valuable as reference material whenever you must select a steel for any specific application as it will guide you in the selection of the proper type.

When you complete the study of this assignment, answer all questions in Progress Quiz 23.

PROGRESS QUIZ 23

THE PURPOSE OF THIS QUIZ: To see how well you understand the different classifications of steel and what they are used for

1. A D The product which results from annealing white cast iron is called malleable iron. (See textbook, page 452.)

2. Steel containing low percentages of nickel, tungsten, or chromium is classified as a plain________________steel. (Page 452)

3. A D When the carbon content of cast iron is reduced to less than two per cent, the resultant metal is called steel. (Page 452)

4. A D Steels containing more than .50 per cent carbon are classified as high carbon steel. (Page 453)

(OVER)

5. The element in steel which determines whether or not steel will harden when heated to its critical temperature and then quenched in oil or water is________________. (Page 453)

6. A distinguishing feature of__________-____________steel is its rough and scaly surface. (Page 454)

7. The grain structure and mechanical properties of____________castings can be greatly improved by annealing or normalizing. (Page 455)

8. Steels containing high percentages of elements other than carbon are classified as_____________steels. (Page 455)

9. High carbon tool steels are alloyed with_________________to increase their resistance to shock. (Page 456)

10. Large amounts of silicon when added to steel will increase the _________________properties of the steel. (Page 456)

11. A D Steels containing high percentages of nickel are non-magnetic. (Page 456)

12. A D A good choice of steel for making cutting tools would be an SAE 5140 steel. (Page 462)

13. The color coding used to identify an SAE 1040 steel is_______________. (Page 463)

(END OF QUIZ)

SUGGESTION: Check your answers to quiz questions by using the Answer Key in the back of the Study Guide. Review all points missed by referring to page references given after each question. Begin your study of Assignment 2.

MACHINE SHOP OPERATIONS AND SETUPS

ASSIGNMENT 2 and PROGRESS QUIZ 24

SUBJECT Heat Treating: Methods, Equipment, and Hardness Testing

TEXTBOOK PAGES TO STUDY 468 to 487

"YOUR AIM" for this assignment is to learn how qualities such as toughness, hardness, and wear resistance can be provided in steel through proper heat-treating practices.

SOME SUGGESTIONS

As you study the various methods of heat-treating steel, which is done to make the steel better suited structurally and physically for some specific application, try to remember the procedures which must be followed in each process. On pages 470 to 473 where hardening of steel is discussed, note that steel can be hardened by several different methods and that the type of steel to be hardened is what will determine the method you must use. Tempering or drawing and the hardening of steels are the two heat-treating processes which must be thoroughly understood by the machinist or tool and die maker. Regardless of how well cutting tools or the parts of a die are machined, these parts will fail in normal use unless they have been properly hardened and drawn. Skill and knowledge of proper heat-treating practices is an important part of all machine shop work. Keep this in mind as you study this assignment.

When you complete the study of this assignment, answer the questions in Progress Quiz 24.

PROGRESS QUIZ 24

THE PURPOSE OF THIS QUIZ: To see how well you understand the heat-treating processes and the methods used in hardness testing as discussed in this assignment

1. The grain structure of steel is composed of__________and______________. (See textbook, page 469.)

2. The steels used in cutting tools of high quality have a______________grain structure. (Page 469)

3. Steel hardens when it is heated and quenched because of the changes which occur in the______________structure. (Page 469)

(OVER)

4. The temperature to which steel must be heated so that it will harden when quenched is called the _______________ temperature. (Page 470)

5. A D The carbon content of steel determines the temperature to which steel must be heated for hardening. (Page 470)

6. A D A hydrometer is an instrument used to check the temperature of a heat-treating furnace. (Page 470)

7. The two processes used in case-hardening steel are _________________ and _________________. (Page 472)

8. A D Low carbon steel can be hardened only by carburizing or cyaniding. (Page 472)

9. A D The depth of case hardness obtained by cyaniding is greater than that obtained by carburizing. (Pages 472-473)

10. A D Safety glasses or a face shield should be worn when hardening parts by cyaniding. (Page 473)

11. All parts which have been hardened must be _______________ to relieve the hardening strains and to increase the toughness of the part. (Page 475)

12. A D Steel parts which are tempered at 500 degrees Fahrenheit will not retain as much hardness as those parts which are tempered at 400 degrees. (Pages 474-475)

13. A D To anneal hard steel so that it can be machined, the steel must be heated above its critical temperature and then cooled slowly. (Page 475)

14. A D To normalize steel, it must be heated above its critical temperature and quenched in oil. (Page 475)

15. A D Carbon steels are hardened by heating the steel to its critical temperature and then quenching it in water. (Page 477)

(END OF QUIZ)

SUGGESTION: Check your answers to this quiz by using the Answer Key in the back of the Study Guide. Review all points missed by referring to page references given after each question. Answer all questions in Memory Jogger 3.

MACHINE SHOP OPERATIONS AND SETUPS

MEMORY JOGGER 3

THE PURPOSE OF THIS QUIZ: To see how well you can recall the important points studied in Chapters 9, 10, 12, 13, and 14 of your textbook

1. A D When the workpiece in a milling operation is fed in a direction opposite the rotation of the cutter, the operation is referred to as conventional milling. (See textbook, page 280.)

2. End-milling operations can be performed on a horizontal milling machine equipped with a ________________ milling attachment. (Page 299)

3. A D To mill twenty teeth in a gear, the index crank of the dividing head must be turned two complete turns, to rotate the work into position for milling each tooth. (Page 301)

4. The operation of milling two sides of a workpiece simultaneously is called ________________ milling. (Page 311)

5. A D A climb milling cut should never be taken when milling cast iron parts as the hard surface scale will dull the cutter. (Page 320)

6. Two important factors which must be considered before attempting to use carbide-tipped milling cutters in preference to high-speed cutters in performing a milling operation, are the ________________ rating and the ____________________ of the machine. (Pages 321-322)

7. A D The only factor which must be considered to determine the correct speed to use in milling is the hardness of the metal to be machined. (Page 323)

8. A D For best results when taking a finishing cut on a milling machine, increase the cutter speed and the feed rate. (Page 326)

9. To find the diametral pitch of a gear, divide the number of teeth in a gear by the ______________ ________________. (Page 344)

10. A D The diametral pitch of a gear having 36 teeth and a pitch diameter of 9 inches is 4. (Page 344)

11. A ________________ ____________ must be used in gear milling operations to obtain the even spacing of the gear teeth. (Page 346)

(OVER)

12. When milling the teeth of a spur gear, the gear blank is mounted on a ________________held between dividing head centers. (Page 347)

13. A D A universal milling machine is not needed when milling spur gear teeth. This applies only to the milling of helical gears. (Page 349)

14. A D When milling a spur gear, the gear blank is rotated by turning the index crank of the dividing head; but when milling helical gears the blank rotates automatically when the power table feed is engaged. (Page 349)

15. A D The table of the milling machine must be set at an angle of 30° to the column of the machine in all helical milling operations. (Page 350)

16. A________________dividing head must be used in all helical gear milling operations to permit gearing the head to the table lead screw. (Page 351)

17. A D On a vertical-spindle surface grinder equipped with a rotating table, the table and grinding wheel rotate in opposite directions. (Page 412, Fig. 1)

18. A D To increase the depth of cut in surface grinding operations, the machine table is raised. (Page 413)

19. A D The operation of sharpening a grinding wheel is called wheel dressing. (Page 420)

20. A D In centerless grinding operations, the work is rotated by the regulating wheel. (Page 428)

21. A D To perform internal grinding on a centerless grinder, the work is rotated and held by a set of rollers. (Page 431)

22. A D An aluminum-oxide is a better wheel to use when grinding any of the hard steels, than the silicon carbide wheel. (Page 443)

23. A D Steels whose carbon content ranges from .05 to .30 per cent are classified as low-carbon steels. (Page 453)

24. A D Because of its dark scaled surface, hot-rolled steel is distinguished from cold-rolled steel. (Page 454)

25. A D It is necessary to temper all parts after hardening, to relieve hardening strains. (Page 475)

(END OF QUIZ)

SUGGESTION: Check your answers to this quiz by using the Answer Key in the back of the Study Guide. Review all points missed. Answer all questions in Examination 9.

CUT OFF HERE

MACHINE SHOP OPERATIONS AND SETUPS **Examination 9**

TOPIC: Steel and Its Alloys, Heat Treating

Study pages 450 to 487 in the textbook

Student's Name ______________________ Student Number ____________

Street ____________ City ____________ State ____________ Zip Code ____

1. Steel is made from cast iron by removing all excess —

   a. carbon
   b. silicon
   c. sulphur

2. Under the general classification of cast irons, the type most suitable for withstanding shock and vibration without danger of cracking is—

   a. chilled cast iron
   b. gray cast iron
   c. malleable iron

3. The most important factor relative to the physical properties of steel is—

   a. carbon
   b. silicon
   c. manganese

4. Machine steel or mild steel are terms commonly applied to steels containing—

   a. more than .40% carbon
   b. less than .30% carbon
   c. more than .50% carbon

5. <u>Two</u> alloying elements of steel commonly used as purifiers or cleansing agents are—

   a. manganese and silicon
   b. vanadium and chromium
   c. molybdenum and nickel

(OVER)

6. An alloying element commonly used in steel to reduce its magnetic properties is—

   a. tungsten
   b. silicon
   c. nickel

7. Hot work steels are alloy steels having a relatively—

   a. low carbon content
   b. high carbon content
   c. low sulphur content

8. The element most prominent in high speed steel, a form of alloy steel, is—

   a. carbon
   b. tungsten
   c. chromium

9. A special characteristic of an SAE 1112 steel is its—

   a. ability to resist shock
   b. excellent machining qualities
   c. ability to resist corrosion

10. Under the Society of Automotive Engineers coding system, an SAE 3140 steel would be classified as a—

    a. manganese steel
    b. chrome-vanadium steel
    c. nickel chromium steel

11. Under the AISI method of classifying steel an A-3115 steel is classified as an—

    a. open-hearth carbon steel
    b. open-hearth alloy steel
    c. electric furnace alloy steel

12. Under the color code for marking steel bars so they can be readily identified as to type of steel, a free cutting steel would have the end of the bar painted—

    a. yellow
    b. green
    c. brown

Student's Name________________________ Student Number______________

13. A test commonly applied to steel of unknown quality for identification purposes is the—

    a. acid-etch test
    b. spark test
    c. fracture test

14. The element in steel which directly affects the critical temperature of the metal to be heat-treated is—

    a. sulphur
    b. phosphorous
    c. carbon

15. The rate of heat transfer, or the ability to heat metal most rapidly, is obtained in—

    a. molten baths
    b. a fuel-fired furnace
    c. electric furnaces

16. High alloy steels must be heated slowly and uniformly for hardening, to avoid—

    a. scaling
    b. shrinkage
    c. warpage

17. Overheating high alloy steels when pack hardening must be avoided to prevent—

    a. low hardness and shrinkage
    b. extreme hardness and brittleness
    c. warpage

18. Localized hardening which refers to objects which require that only a small selected portion of the object be hardened, can be accomplished only by—

    a. flame and induction hardening
    b. pack hardening
    c. cyaniding

CUT OFF HERE

(OVER)

19. Case hardening is the only method suitable for hardening—

    a. high alloy steel
    b. high carbon steel
    c. low carbon steel

20. A D Complete penetration of hardness cannot be obtained in parts made from low carbon steel.

21. When hardening steel by the carburizing process and the steel has been heated to the correct temperature for the correct amount of time, the furnace is shut off and the steel is—

    a. removed and quenched in water
    b. left in the furnace to cool
    c. removed and cooled in air

22. The greatest depth of case or the extent to which hardness penetrates in the cyaniding process for hardening steel is—

    a. .015
    b. .005
    c. .025

23. The depth of the case or the extent to which hardness would penetrate in steel hardened by the cyaniding process if left in the bath for seven minutes, would range from—

    a. .004 to .006
    b. .015 to .020
    c. .007 to .0105

24. Hardening strains created in steel after it has been heated and quenched must be removed by—

    a. tempering
    b. annealing
    c. normalizing

25. The maximum rate at which steel should be cooled in a furnace from its annealing temperature is—

    a. 90° Fahrenheit per hour
    b. 75° Fahrenheit per hour
    c. 50° Fahrenheit per hour

Student's Name______________________ Student Number______________

26. The heat-treating process used to soften hard alloy and tool steels so that they can be more easily machined is called—

    a. carburizing
    b. annealing
    c. normalizing

27. Because of the manner in which carbon steels are quenched from the hardening heat, they are classified as—

    a. air-hardening steels
    b. oil-hardening steels
    c. water-hardening steels

28. A D The temperature of the water used in quenching water-hardening steels from the hardening heat, can range from 40 to 100 degrees Fahrenheit.

29. A D High alloy steels are not hardened by quenching. They are cooled from the hardening heat in still air.

30. Proper control of the________________in a heat-treating furnace is necessary to prevent excessive scaling of parts being hardened.

31. The atmosphere of a fuel-fired, heating-treating furnace is controlled by regulating the__________-____________ratio.

32. The temperature of furnaces used for heat-treating steel is measured and controlled by an instrument called a—

    a. pyrometer
    b. thermometer
    c. hydrometer

(END OF EXAMINATION)

—CUT OFF HERE—

SECTION 10 — Study Guide for MACHINE SHOP OPERATIONS AND SETUPS

TEXTBOOK CHAPTERS TO STUDY

Chapters 15, 16, and 17 — Machinability, Numerical Control, Electrical Energy Processes

CONTENTS OF SECTION 10

Assignment 1 and Progress Quiz 25 — Study pages 488 to 509, the methods used and the factors to consider in determining the machinability of metals.

Assignment 2 and Progress Quiz 26 — Study pages 510 to 526, the automatic operation of machine tools by numerical control.

Assignment 3 and Progress Quiz 27 — Study pages 527 to 544, the machining of parts by electrical energy processes.

EXAMINATION 10 — Based on pages 488 to 544 of your textbook

WHAT THE STUDY OF SECTION 10 WILL DO FOR YOU

The subject material in this final section of the textbook is quite varied as it covers three separate but short chapters. The material in Chapter 15 deals with the subject of what factors determine the machinability of specific metals. Unless you learn what these factors are, you may experience extreme difficulty when machining metals classified as difficult to machine. Knowing what must be done to obtain the best results when machining these metals is very important and will help you to boost machining performance and thereby reduce machining costs.

In Chapter 16 you will read and learn about how a recently developed system called Numerical Control has revolutionized many manufacturing operations which include the machining of metal. This system, when applied to the conventional machine tools studied in earlier textbook chapters, is capable of directing and commanding these machine tools to perform all operations automatically from instructions coded on a punched tape. To keep abreast of the times, it is important that the general machinist and tool and die maker understand the principals of this system and exactly how it functions. The same

(OVER)

holds true for the recent development of machine tools, which use electrical energy to accomplish the machining of metals and materials as discussed in Chapter 17 of your textbook. It will be to your advantage to have a working knowledge of these new developments in machining technology as these machine tools are capable of performing some operations more economically and with better results than can be obtained by conventional machining. This is the knowledge you will acquire in the study of this section of the textbook.

Begin your study of Assignment 1.

MACHINE SHOP OPERATIONS AND SETUPS

ASSIGNMENT 1 and PROGRESS QUIZ 25

SUBJECT Machinability: Variables and Ratings

TEXTBOOK PAGES TO STUDY 488 to 509

"YOUR AIM" for this assignment is to acquire understanding of why certain metals can be easily machined with a fine surface finish under some conditions but may be difficult to machine when these same conditions are varied.

SOME SUGGESTIONS

As you study this assignment, try to develop an understanding of what factors are used to determine the machinability ratings of the various steels shown in tables 1 and 2 on pages 507 and 508 of your textbook. This is very important in the machining of metals classified as being difficult to machine, because you can control most of the factors used in obtaining these ratings and thereby increase tool life and obtain a high quality surface finish when machining these "difficult-to-machine" steels.

When you complete the study of this assignment, answer the questions in Progress Quiz 25.

PROGRESS QUIZ 25

THE PURPOSE OF THIS QUIZ: To see how well you understand the factors considered in determining the machinability rating of various metals

1. Two factors which can affect the efficiency of any machining operation are the machine adjustments made for____________and____________. (Page 488)

2. The cutting efficiency of a tool at a given speed can be determined by observing the______________of the chips produced. (Page 489)

3. A D When machining work with carbide tools, a blue or gray chip is an indication that a correct cutting speed is being used. (Page 489)

4. A D Metal which can be machined at high speeds would be rated as having a high machinability. (Pages 492-493)

(OVER)

5. A D The rate of tool wear and the finish quality of a machined surface are both affected by the type of chip produced. (Page 493)

6. Discontinuous chips are those normally obtained when machining brittle metals such as____________ ____________. (Page 493)

7. A D Continuous chips are those normally produced when machining soft ductile metals. (Page 494)

8. An important factor to be considered in the selection of a cutting tool material is the type of________________to be machined. (Page 496)

9. The hardness of carbon tool steels will be increased when alloyed with ________________and vanadium. (Page 497)

10. The type of wear which occurs on the face of tool bits is referred to as ________________wear. (Page 498)

11. A D Maximum cutting speeds are only obtainable when machining most metals if a cutting fluid is used. (Page 506)

12. The tool steels easiest to machine are those classified as____________ tool steels. (Page 509)

(END OF QUIZ)

SUGGESTION: Check your answers to quiz questions using the Answer Key in the back of the Study Guide. Review all points missed, then begin your study of Assignment 2.

MACHINE SHOP OPERATIONS AND SETUPS

ASSIGNMENT 2 and PROGRESS QUIZ 26

SUBJECT Numerical Control: Point-to-Point, and Continuous Path Systems

TEXTBOOK PAGES TO STUDY 510 to 526

"YOUR AIM" for this assignment is to learn how the complete control and operation of any machine tool is performed automatically by numerical control.

SOME SUGGESTIONS

To understand how a numerically controlled machine tool can be directed to perform all functional movements needed in machining a part from start to finish without the manual adjustments made by a machine operator, as in the conventional operation of a machine tool, you must understand the rectangular coordinate system of measuring as shown in Fig. 1 on page 514 of your textbook. As you study Fig. 1, try to recall the basic movements of each machine tool you read about in earlier chapters of your textbook so you can relate each movement to one of the three axes shown in Fig. 1. The horizontal milling machine is a good example. The table of this machine can be moved longitudinally in a horizontal plane, at right angles to this longitudinal movement in a horizontal plane (cross feed), and it can be raised or lowered in a vertical plane. Keeping these movements in mind as you study Fig. 1 on page 514 of the textbook, you can see that the longitudinal table movement can be considered to correspond to any movement in the X-axes. Any cross feed table movement needed can be considered to correspond to movement in the Y-axes of Fig. 1, and any vertical table movement for depth of cut can be considered to correspond to movement in the Z- or vertical axes of Fig. 1 on page 514 of your textbook. The adjustments or major movements needed on each machine tool studied in earlier chapters of the textbook can be related to the X-Y- and Z-axes of Fig. 1 on page 514 of the textbook in this same manner. An understanding of Fig. 1 and how it

(OVER)

relates to machine tool adjustments or movements will help you to understand the discussion of all numerical control principles as explained in the remainder of Chapter 16, beginning on page 510 of your textbook.

After you complete the study of this assignment, answer all questions in Progress Quiz 26.

MACHINE SHOP OPERATIONS AND SETUPS

PROGRESS QUIZ 26

THE PURPOSE OF THIS QUIZ: To see how well you understand the basic principles of Numerical Control and how it relates to the automatic operation of all machine tools

1. The two systems of numerical control are____________-to-____________ and the continuous path system. (See textbook, page 513.)

2. The machine tool to which the system of point-to-point numerical control applies is the____________ ____________. (Page 513)

3. The numerical control system which applies to a milling machine is called the____________ ____________system. (Page 513)

4. A D A computer must be used to prepare instructions for all numerically controlled machine tools. (Page 514)

5. The man who is responsible for listing the sequence of operations or machine movements needed to complete the machining of a part on a numerically controlled machine tool is called a____________. (Page 521)

6. A D In some numerical control programs, the machine operator may be required to perform tool changes and other operations manually. (Page 521)

7. A D To be a part programmer, you must have complete knowledge of the setups, the tooling, and the operation of all conventional machine tools. (Page 525)

8. A D More machining skill is needed to operate a conventional machine tool than is needed to operate a numerically controlled machine tool. (Page 525)

9. A D Skilled machinists and toolmakers with a good knowledge of shop mathematics are frequently used as part programmers. (Page 525)

10. A D A knowledge of electronics is essential for anyone wanting to learn how to service or maintain numerically controlled machine tools. (Page 525)

(END OF QUIZ)

SUGGESTION: Check your answers to quiz questions by using the Answer Key in the back of the Study Guide. Review all points missed, then begin your study of Assignment 3.

MACHINE SHOP OPERATIONS AND SETUPS

ASSIGNMENT 3 and PROGRESS QUIZ 27

SUBJECT Electrical Energy Processes: Electro-Discharge, Electro-Chemical, Ultrasonic, Magnetic Pulse Forming, and Electrolytic Grinding

TEXTBOOK PAGES TO STUDY 529 to 544

"YOUR AIM" for this assignment is to learn about the newly developed electrical energy machining processes used to machine the hard exotic metals which are difficult or impossible to machine by conventional methods.

SOME SUGGESTIONS

In the study of this assignment, you will read and learn about the five newly developed electrical energy machines and machining processes. As you study each process, learn the major elements or parts of each machine tool and what function each element performs, what type of cutting tool or electrode must be used, and what determines the choice of electrode material in each operation. Also note the type of work each process is best suited to perform and the type of electrolyte bath used in each process.

When you complete the study of this assignment, answer all questions in Progress Quiz 27.

PROGRESS QUIZ 27

THE PURPOSE OF THIS QUIZ: To see how well you understand the various electrical energy machining processes and the machines used.

1. A D Hardened alloy steels which cannot be machined with a conventional cutting tool can be machined by Electro-Discharge Machining. (See textbook, page 527.)

2. A D In the EDM process, only the workpiece is submerged in a dielectric fluid. The electrode must be kept dry. (Page 527)

3. A D The rate of machining in the EDM process can be varied by increasing or decreasing the number of electrical discharges per second. (Page 528)

4. The cutting tool used in the EDM process is called an_________________. (Page 529)

(OVER)

5. The two electrical circuits used in the power supply unit of EDM machines are the_______________type and the_______________type. (Page 531)

6. The machining process where the metal of a workpiece is dissolved into an electrolyte solution is called_______________-_______________ machining. (Pages 531-532)

7. A D The abrasives used in ultrasonic machining are the same as those used in conventional grinding wheels. (Page 537)

8. A machine which is used to form aluminum tubing electromagnetically is called a____________________machine. (Pages 539-540)

9. A D In the electrolytic grinding process, most of the metal is removed from the workpiece by deplating rather than by grinding. (Page 542)

10. The wheel head of an electrolytic grinder must be__________________ from the wheel-head housing and the rest of the machine. (Page 543)

(END OF QUIZ)

SUGGESTION: Check your answers to this quiz by using the Answer Key in the back of the Study Guide. Review all points missed. Answer all questions in Examination 10.

MACHINE SHOP OPERATIONS AND SETUPS

Examination 10

TOPIC: Machinability, Numerical Control, Electrical Energy Machining Processes

Study pages 488 to 544 in the textbook

Student's Name____________________ Student Number__________

Street__________ City__________ State__________ Zip Code_____

1. A skilled machinist can determine if he is using the most efficient cutting speed in a machining operation by the color of the chip produced. What should be the color of the chip produced when using a ceramic tool bit?

   a. brown
   b. purple
   c. gray

2. An efficient cutting speed is indicated, if carbide tipped cutting tools are used, when the color of the chip produced is__________or__________.

3. A D The use of a proper cutting fluid for the metal being machined will improve the machinability of the metal.

4. A D The surface finish produced is not an important factor to be considered in determining the machinability rating of a metal.

5. The form of chip produced when machining a hard brittle metal like cast iron is the—

   a. continuous chip
   b. discontinuous chip
   c. continuous chip with built-up edge

6. A D A cutting tool ground so that it forms a continuous type of chip will produce the best surface finish on a machined part.

7. A D A 1025 SAE hot-rolled steel has a lower machinability rating than a 1025 SAE cold-drawn steel.

8. A D Hot-rolled, low-carbon steels which are relatively soft are more difficult to machine than some of the much harder alloy steels.

(OVER)

F4-4

9. Of all tool steels in use, the one with the highest machinability rating is a—

   a. carbon-vanadium tool steel
   b. manganese oil-hardening tool steel
   c. straight carbon tool steel

10. The method by which conventional machine tools are directed to perform all operations automatically from instructions on a roll of tape is called ______________________ __________________.

11. The point-to-point system of numerical control can be applied only to—

   a. conventional drill press or jig boring operations
   b. conventional milling operations
   c. conventional shaper operations

12. The vertical movement of the table on a conventional milling machine is represented in the Cartesian coordinate system as a movement parallel to the—

   a. Z-axes
   b. X-axes
   c. Y-axes

13. The cross feed movement of a conventional milling machine table is represented in the Cartesian coordinate system as a movement parallel to the—

   a. X-axes
   b. Y-axes
   c. Z-axes

14. The man who must prepare a manuscript program to show the sequence of operations to be performed by a numerically controlled machine tool is called a______________________.

15. A D In some numerical control programs, the machine operator may be required to perform tool changes manually.

16. The control unit which commands and guides the movement of numerically controlled machine tools is called a__________ __________________.

17. Programming a contouring cut on a milling machine is accomplished faster when a__________________is used.

Student's Name______________________________Student Number________________

18. A D More machining skill is needed to operate a conventional machine tool than is needed to operate a numerically controlled machine tool.

19. A knowledge of ________________ is very helpful to anyone wanting to learn how to service or maintain numerically controlled machine tools.

20. A D The electrodes used in the ECM process must be made of an electric conducting metal.

21. Both the electrode and the workpiece must be submerged in a dielectric fluid when machining work by the—

    a. USM process
    b. ECM process
    c. EDM process

22. A D The electrodes used in the ECM process differ from those used in the EDM process in that the ECM electrodes must be insulated.

23. In the ECM process, machining is accomplished by feeding an____________ fluid through an electrically charged hollow electrode.

24. The part of an ultrasonic machine tool which causes the tool to vibrate at a rate of 20,000 times per second is called a____________________.

25. The wheel head of an electrolytic grinder must be______________from the wheel head housing and the rest of the machine.

CUT OFF HERE

(END OF EXAMINATION)

MACHINE SHOP OPERATIONS AND SETUPS

Answer Key

Progress Quiz 1

1. tools
2. cube, sphere
3. D
4. measurement
5. files
6. safety glasses
7. rings
8. damaged
9. oil-soaking

Progress Quiz 2

1. 1/64
2. D
3. layout
4. angles
5. rule depth
6. calipers
7. A
8. calipers
9. A

Progress Quiz 3

1. A
2. A
3. .055
4. .010
5. .0005

Progress Quiz 3—Continued

6. .0625
7. .375
8. .625
9. meter
10. A

Progress Quiz 4

1. A
2. A
3. .313
4. .460
5. .772
6. .713
7. .749
8. .644
9. .3633
10. .2647

Progress Quiz 5

1. A
2. A
3. .01
4. .025
5. A
6. A
7. vernier, protractor
8. 11°35'

Progress Quiz 6

1. D
2. A
3. A
4. D
5. file card
6. safety glasses
7. tap
8. A
9. die
10. reamer
11. pitch
12. A
13. A

Progress Quiz 7

1. machined
2. surface plate
3. scriber
4. holes
5. divider
6. keyseat
7. D
8. A
9. angle plate
10. combination square

Progress Quiz 8

1. hack
2. band

Progress Quiz 8—Continued

3. vertical
4. stops
5. tension
6. annealed
7. pitch
8. 1/64
9. prick punch
10. safety glasses

Progress Quiz 9

1. D
2. A
3. D
4. A
5. cutting edges
6. clearance
7. flutes
8. margins
9. A
10. numbers, letters
11. D
12. A
13. A

Progress Quiz 10

1. A
2. A
3. aluminum
4. A

Progress Quiz 10—Continued

5. angle, length
6. countersink
7. core
8. reamer
9. A
10. A
11. A
12. A
13. A

Progress Quiz 11

1. D
2. A
3. A
4. V block
5. A
6. countersinking
7. counterboring
8. A
9. D
10. goggles

Memory Jogger 1

1. A
2. A
3. .625
4. .250
5. .025
6. A

Memory Jogger 1—Continued

7. vernier
8. A
9. A
10. vernier bevel protractor
11. A
12. D
13. tap
14. die
15. A
16. layout
17. divider
18. horizontal
19. vertical
20. table
21. tension
22. annealed
23. pitch
24. guides
25. safety
26. gloves

Progress Quiz 12

1. A
2. A
3. headstock
4. tailstock
5. A
6. A
7. wooden cradle

Progress Quiz 12—Continued

8. faceplate
9. gibs
10. A

Progress Quiz 13

1. side rake
2. back rake
3. clearance
4. A
5. center
6. side
7. A
8. fine, high
9. necking
10. tungsten

Progress Quiz 14

1. D
2. A
3. A
4. dial indicator
5. A
6. faceplate
7. length
8. headstock, tailstock
9. above
10. micrometer
11. parallel
12. A

Progress Quiz 14—Continued

13. carriage
14. A

Progress Quiz 15

1. A
2. D
3. Unified Standard
4. A
5. center gage
6. D
7. A
8. A
9. A
10. A
11. Acme
12. square

Progress Quiz 16

1. D
2. tailstock or dead
3. A
4. A
5. D
6. A
7. A
8. A
9. boring
10. micrometer
11. A

Progress Quiz 17

1. collet
2. saddle
3. bar turner
4. taper attachment
5. spade
6. micrometer
7. floating
8. collapsible
9. screw
10. cams
11. cutting-off
12. multiple-spindle
13. spindle
14. .002

Progress Quiz 18

1. D
2. micrometer
3. A
4. hydraulic
5. A
6. parallels
7. cylindrical
8. length
9. D
10. indexing
11. A

Memory Jogger 2

1. independent
2. lathe dog
3. mandrel
4. rake
5. clearance
6. diamond
7. dial indicator
8. b
9. center
10. compound
11. universal
12. vertical
13. parallels
14. c
15. indexing
16. chucking
17. saddle
18. bar turner
19. taper attachment
20. spade
21. collapsible
22. screw
23. cam
24. vertical
25. multiple-spindle
26. .002

Progress Quiz 19

1. A
2. arbor support
3. D
4. spindle
5. knee
6. micrometer
7. A
8. A
9. dividing head
10. A
11. A
12. A
13. A
14. A
15. A
16. 40
17. A
18. D

Progress Quiz 20

1. A
2. A
3. A
4. D
5. side
6. face
7. A
8. longitudinal
9. A

Progress Quiz 20—Continued

10. A
11. A

Progress Quiz 21

1. A
2. shell-end
3. diameter
4. dividing head
5. A
6. A
7. A
8. dividing head
9. universal
10. A
11. addendum

Progress Quiz 22

1. surface
2. wheelhead
3. magnetic
4. D
5. dressing
6. plunge
7. A
8. D
9. D
10. A

Progress Quiz 23

1. A
2. carbon
3. A
4. A
5. carbon
6. hot-rolled
7. steel
8. alloy
9. vanadium
10. magnetic
11. A
12. A
13. green

Progress Quiz 24

1. iron, carbon
2. fine
3. grain
4. critical
5. A
6. D
7. carburizing, cyaniding
8. A
9. D
10. A
11. tempered
12. A
13. A
14. D
15. A

Memory Jogger 3

1. A
2. vertical
3. A
4. straddle
5. A
6. horsepower, rigidity
7. D
8. D
9. pitch diameter
10. A
11. dividing head
12. mandrel
13. A
14. A
15. D
16. universal
17. A
18. D
19. A
20. A
21. A
22. A
23. A
24. A
25. A

Progress Quiz 25

1. speeds, feeds
2. color
3. A
4. D
5. A
6. cast iron
7. A
8. material
9. chromium
10. cratering
11. A
12. carbon

Progress Quiz 26

1. point-to-point
2. drill press
3. continuous path
4. D
5. programmer
6. A
7. A
8. A
9. A
10. A

Progress Quiz 27

1. A
2. D
3. A

Progress Quiz 27—Continued

4. electrode
5. pulse, relaxation
6. electro-chemical
7. A
8. magneform
9. A
10. insulated